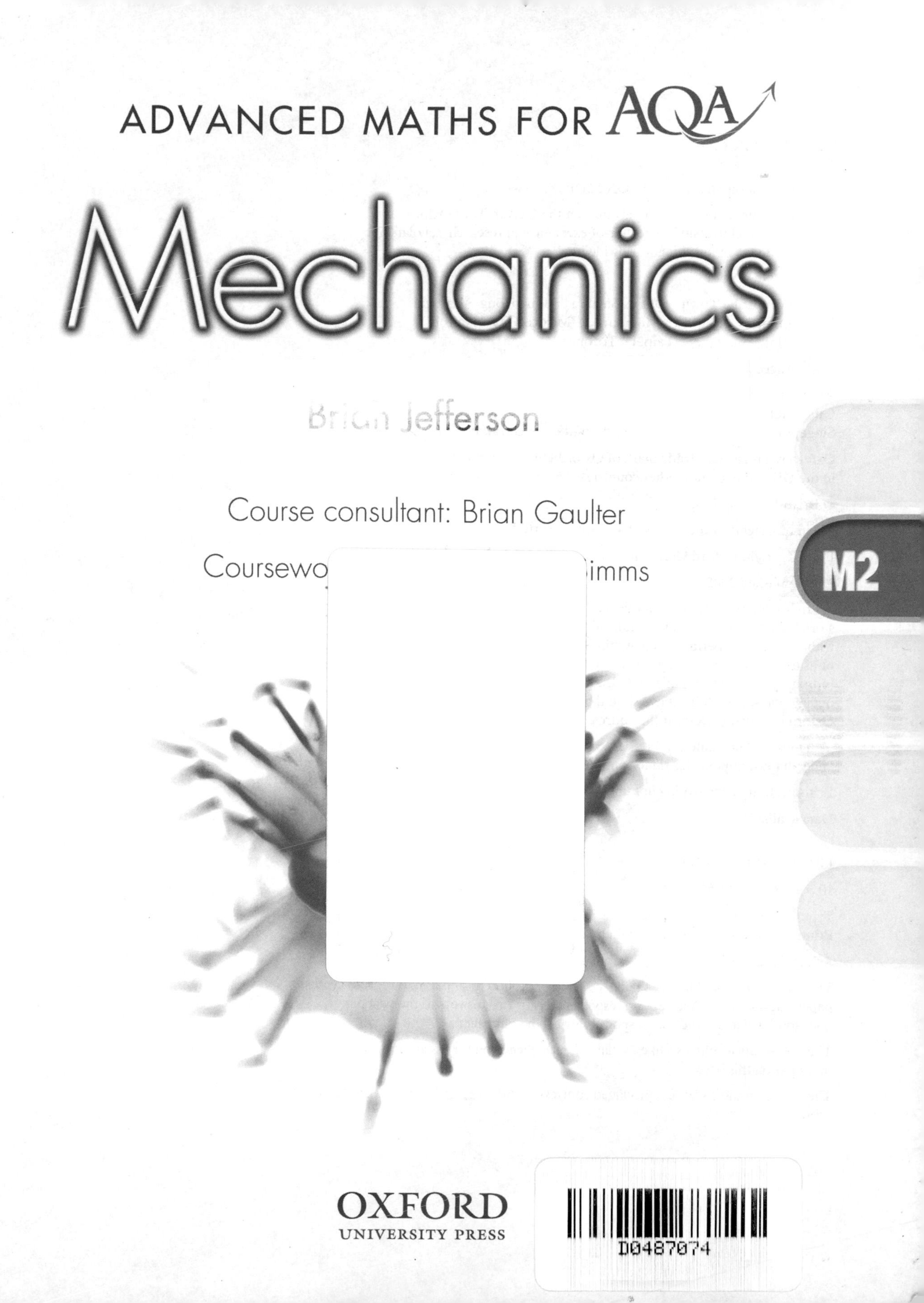

ADVANCED MATHS FOR AQA

Mechanics

Brian Jefferson

Course consultant: Brian Gaulter

Coursewo... ...imms

M2

OXFORD
UNIVERSITY PRESS

OXFORD
UNIVERSITY PRESS

Great Clarendon Street, Oxford OX2 6DP

Oxford University Press is a department of the University of Oxford.
It furthers the University's objective of excellence in research, scholarship,
and education by publishing worldwide in

Oxford New York

Auckland Cape Town Dar es Salaam Hong Kong Karachi
Kuala Lumpur Madrid Melbourne Mexico City Nairobi
New Delhi Shanghai Taipei Toronto

With offices in

Argentina Austria Brazil Chile Czech Republic France Greece
Guatemala Hungary Italy Japan South Korea Poland Portugal
Singapore Switzerland Thailand Turkey Ukraine Vietnam

First published 2005

British Library Cataloguing in Publication Data

Data available

ISBN: 978 019 914989 6

10 9 8 7 6 5 4 3 2

Typeset by Tech-Set Ltd, Gateshead, Tyne and Wear
Printed and bound in Great Britain by Bell and Bain.

Acknowledgements

The publishers would like to thank AQA for their kind permission to reproduce past paper questions. AQA accept no responsibility for the answers to the past paper questions, which are the sole responsibility of the publishers.

The publishers would also like to thank James Nicholson for his authorative guidance in preparing this book.

The image on the cover is reproduced courtesy of Adam Hart-Davis/Science Photo Library.

About this book

This Advanced level book is designed to help you to get your best possible grade in the AQA MM2 (A and B) module for first examination in 2006. This module can contribute to an award in GCE AS level Mathematics or A level Mathematics.

Each chapter starts with an overview of what you are going to learn and a list of what you should already know. The 'Before you start' section contains 'Check in' questions, which will help to prepare you for the topics in the chapter.

You should know how to ...	Check in
1 Manipulate algebraic equations.	**1** Find v if $\frac{1}{2}mv^2 + mgh = 3mgh$

M2

Key information is highlighted in the text so you can see the facts you need to learn.

$$\text{velocity} = \frac{\text{change of displacement}}{\text{time taken to change}}$$

Worked examples showing the key skills and techniques you need to develop are shown in boxes. Also hint boxes show tips and reminders you may find useful.

Example 12

A crane lifts a load of 50 kg to a height of 12 m in a time of 20 s. Find the power required.

The work done against gravity is $50g \times 12 = 5880$ J

Assuming that you can neglect any resistance forces, this is the work done by the crane. Therefore, you have

Time taken to lift load $= 20$ s
Rate of working of crane $= 5880 \div 20 = 294$ W

So, the power required is 294 W.

rate = work done ÷ time taken

The questions are carefully graded, with lots of basic practice provided at the beginning of each exercise.

At the end of an exercise, you will sometimes find underlined questions.

10 Find the work done by a force $\mathbf{F} = (5\mathbf{i} + 3\mathbf{j})$ N whose point of application undergoes a displacement $\mathbf{s} = (3\mathbf{i} + 7\mathbf{j})$ m.

These are optional questions that go beyond the requirements of the specification and are provided as a challenge.

At the end of each chapter, there is a summary. The 'You should now know' section is useful as a quick revision guide, and each 'Check out' question identifies important techniques that you should remember.

You should know how to ...	Check out
1 Find the position, velocity and acceleration of a particle moving with variable acceleration.	**1** A particle starts from rest at time $t = 0$. Its acceleration is $a = 3t^2 + e^{2t}$. Find its velocity and position at time t.

Following the summary, you will find a revision exercise with past paper questions from AQA. These will enable you to become familiar with the style of questions you will see in the exam.

A special feature of the text is the reference to a number of spreadsheets, used to analyse the data from suggested experiments or to explore the implications of certain models. These can be downloaded from the Oxford University Press website:

(*http://www.oup.co.uk/secondary/mechanics*)

Practice Papers, written by a senior examiner, will directly help you to prepare for your exams.

The book also contains a chapter devoted to coursework guidance for students taking Unit MM2A. Written by a senior moderator, this section allows you to understand the requirements and to fully prepare for the coursework component.

At the end of the book, you will find numerical answers and a list of formulae you need to learn.

Contents

1 Calculus in kinematics

This chapter will show you how to

- Use calculus to find the position, velocity and acceleration of a particle moving with variable acceleration along a straight line
- Use calculus to find the position, velocity and acceleration vectors of a particle moving in two- or three-dimensional space
- Apply Newton's laws of motion to situations involving variable acceleration

Before you start

You should know how to ...	Check in
1 Differentiate polynomials and algebraic, exponential, logarithmic and trigonometric functions.	**1** Find $\frac{dy}{dx}$ where: a) $y = 3x^3 + 2x$ b) $y = x^{\frac{5}{2}} + \frac{7}{\sqrt{x}} + \frac{3}{x^2}$ c) $y = \ln(2x + 1)$ d) $y = e^{5x}$ e) $y = \sin 4x$ f) $y = x^3 \cos 2x$
2 Integrate polynomials and algebraic, exponential and trigonometric functions.	**2** Find: a) $\int (x^4 + 2x^3)\,dx$ b) $\int \frac{2x+5}{\sqrt{x}}\,dx$ c) $\int e^{5x}\,dx$ d) $\int \sin 3x\,dx$
3 Integrate expressions using substitution.	**3** Find: a) $\int x(x+5)^3\,dx$ b) $\int x^2 e^{x^3}\,dx$ (Hint: let $u = x^3$)
4 Evaluate definite integrals.	**4** Evaluate $\int_1^3 (x^2 + 3x)\,dx$
5 Manipulate vectors.	**5** Given $\mathbf{r} = 3\mathbf{i} + 7\mathbf{j}$ and $\mathbf{s} = 7\mathbf{i} - 5\mathbf{j}$, find: a) the magnitude of $\mathbf{r} - \mathbf{s}$ b) $2\mathbf{r} + 3\mathbf{s}$
6 Know the constant acceleration equations in vector form.	**6** Write down the four vector equations for motion with constant acceleration.
7 Know Newton's laws of motion.	**7** A body of mass 6 kg is acted upon by forces $(4\mathbf{i} + 7\mathbf{j})$ N and $(2\mathbf{i} + 5\mathbf{j})$ N. Find the acceleration of the body.

1.1 Velocity at an instant

Displacement–time graphs are described in the Mechanics 1 module. When the velocity of a body is constant, the graph is a straight line and the velocity is represented by the gradient of the graph.

$$\text{Gradient} = \frac{\text{Change of displacement}}{\text{Time taken to change}} = \text{Velocity}$$

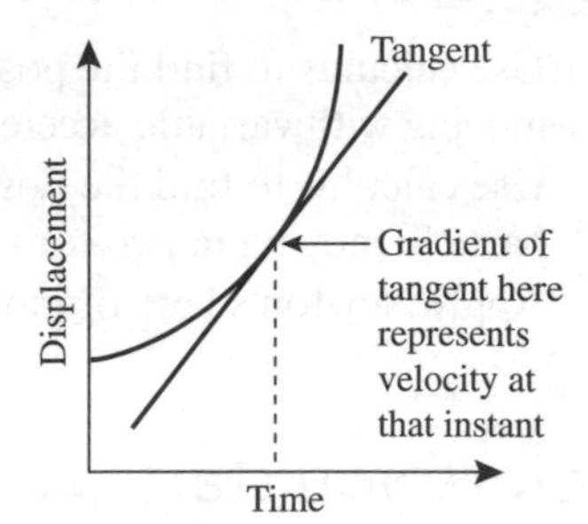

When the displacement–time graph is not linear the gradient changes, but it is still true that the gradient at a point on the graph represents the velocity at that instant.

If you know the equation of a graph, you can often find the gradient by differentiation. Using v for velocity and x for displacement, you write

$$v = \frac{dx}{dt}$$

Any derivative formed by differentiating with respect to time creates a **rate of change**. In this case, velocity is the rate of change of displacement.

Similarly, acceleration is the rate of change of velocity. You can therefore find acceleration by differentiating velocity with respect to time. This is equivalent to differentiating displacement twice.

$$a = \frac{dv}{dt} = \frac{d^2x}{dt^2}$$

Note For differentiation with respect to time there is an alternative notation that puts dots over the dependent variable. You would write

$$v = \dot{x}$$

and $a = \dot{v} = \ddot{x}$

These relationships can also be written in terms of integration.

You should know from Mechanics 1 that the area under a velocity–time graph represents the change in displacement. In general, this area is found by integration. This corresponds to the relationship

$$x = \int v \, dt$$

Similarly, the area under an acceleration–time graph represents the velocity, giving

$$v = \int a \, dt$$

These results enable you to solve kinematics problems in one dimension when the acceleration of the body is variable, provided that it can be expressed in terms of t.

Example 1

A particle moves in a straight line so that at time t seconds its displacement x metres from an origin is given by $x = t^4 - 32t$. Find the position and acceleration of the particle at the moment when it comes instantaneously to rest.

By differentiating, you find expressions for the velocity and acceleration of the particle at time t.

$$v = \frac{dx}{dt} = 4t^3 - 32 \quad \text{and} \quad a = \frac{dv}{dt} = 12t^2$$

The particle is at rest when $v = 0$.

This gives $4t^3 - 32 = 0$

Hence $t^3 = 8$ and so $t = 2$ s

The position of the particle at this moment is $x = 2^4 - 32 \times 2 = -48$ m

and the acceleration is $a = 12 \times 2^2 = 48 \text{ m s}^{-2}$

Example 2

A particle travels along a straight wire starting from a point A. Its velocity is 6 m s^{-1} at time $t = 0$ and its acceleration is given by $a = -6t \text{ m s}^{-2}$.

a) Find its velocity when $t = 1, 2, 3$ seconds.

b) Find its position when $t = 1, 2, 3$ seconds.

c) Find its change of position during the first 3 seconds of its motion.

d) Find how far it has travelled during the first 3 seconds of its motion.

a) To find an expression for velocity, integrate acceleration.

$$v = \int -6t \, dt = -3t^2 + c$$

When $t = 0$, $v = 6$, from which you evaluate $c = 6$.
Hence $v = 6 - 3t^2$.
Substituting $t = 1, 2$ and 3 gives

t	1	2	3
v	3	−6	−21

b) To find an expression for displacement, integrate velocity.

$$x = \int (6 - 3t^2) \, dt = 6t - t^3 + k$$

When $t = 0$, $x = 0$, from which you evaluate $k = 0$.
Hence $x = 6t - t^3$.
Substituting $t = 1, 2$ and 3 gives

t	1	2	3
v	5	4	−9

c) The particle is at A when $t = 0$ and at position -9 when $t = 3$. It has undergone a displacement of -9 m.

d) The diagram illustrates the motion of the particle.

D	A	C	B
$t = 3$	$t = 0$	$t = 2$	$t = 1$
← $v = -21$	$v = 6$ →	← $v = -6$	$v = 3$ →
$x = -9$	$x = 4$	$x = 4$	$x = 5$

The particle starts at A and passes through B when $t = 1$. At that moment, it is still moving to the right at $3\,\text{m s}^{-1}$. During the second second it stops momentarily, before returning through C and A and finally travelling to D. (After this time it continues to travel to the left in the diagram and at an increasing speed.)

To find the total distance travelled, you need to find out where it comes to rest.

At this point, $v = 0$. Therefore

$6 - 3t^2 = 0$ giving $t = \sqrt{2}$ s (or $-\sqrt{2}$, which is inappropriate).

The position of the particle at $t = \sqrt{2}$ is

$$x = 6 \times \sqrt{2} - (\sqrt{2})^3 = 4\sqrt{2}$$

The total distance travelled in the first 3 seconds is therefore

$$(4\sqrt{2} + 4\sqrt{2} + 9) = (9 + 8\sqrt{2})\text{ m}$$

The change of position (displacement) in part c) could have been found as a definite integral $\int_0^3 (6 - 3t^2)\,dt$. This does **not** give the total distance travelled.

An extended model

The next two examples explore two possible models for the motion of a car travelling between two sets of traffic lights. The first examines a simple but flawed model, while the second attempts to devise a more realistic model of the situation.

Example 3

A car is travelling between two sets of traffic lights. Starting from rest at the first set of lights, it accelerates up to a maximum speed before slowing down to a stop at the second set of traffic lights.

The model proposed is that at time t s the displacement x m of the car from its starting point is given by the formula

$$x = \tfrac{1}{45}t^2(45 - t)$$

Describe the motion of the car using this model, and criticise the model.

Expanding the expression for displacement, you have

$$x = t^2 - \tfrac{1}{45}t^3$$

Differentiate to get

$$v = 2t - \tfrac{1}{15}t^2 = \tfrac{1}{15}t(30 - t)$$

Differentiate again to get

$$a = 2 - \tfrac{2}{15}t$$

From these you can deduce that

- ✦ $v = 0$ when $t = 0$ and when $t = 30$. The journey between the lights takes 30 s.
- ✦ $x = 300$ when $t = 30$. The distance between the lights is 300 m.
- ✦ The velocity is a maximum when $a = 0$, which happens when $t = 15$. When $t = 15$, $v = 15$, so the maximum velocity is 15 m s^{-1}.

The maximum velocity occurs just before the car starts to slow down, so the acceleration is 0.

You can illustrate these characteristics of the motion by drawing graphs of acceleration, velocity and displacement against time.

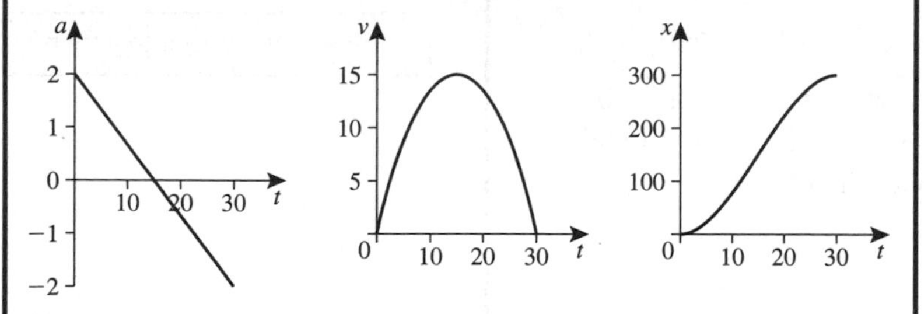

The main problem with the model relates to the beginning and end of the motion. The car is at rest at the start, but suddenly acquires an acceleration of 2 m s^{-2}. This does not correspond to experience, as the acceleration of a car builds gradually from zero. In a similar way, you would expect the deceleration of the car to decrease to zero as the car comes to rest, but the model indicates an acceleration of -2 m s^{-2} at that point (If t were allowed to increase beyond 30 the car would move backwards ever more rapidly).

To make sense, the model would have to include the restriction that $0 \leqslant t \leqslant 30$.

M2

Example 4

Devise a model for the motion of the car in Example 3 with these characteristics:

- ✦ The velocity is zero when $t = 0$ and when $t = 30$.
- ✦ The acceleration is zero when $t = 0$, when $t = 30$ and also when $t = 15$, to correspond to maximum velocity.
- ✦ The distance travelled by the car is 300 m.

Find the maximum velocity as predicted by this model.

The simplest expression for this acceleration is a cubic polynomial of the form

$$a = Kt(t - 15)(t - 30)$$

where K is a constant to be determined.

This satisfies the condition that $a = 0$ when $t = 0$, 15 and 30.

In order to find the velocity, expand the expression for acceleration and integrate.

$$v = \int K(t^3 - 45t^2 + 450t)\,dt = K(\tfrac{1}{4}t^4 - 15t^3 + 225t^2) + c$$

c is the constant of integration

The required initial conditions, $v = 0$ when $t = 0$, give $c = 0$.

Hence, you have $v = K(\tfrac{1}{4}t^4 - 15t^3 + 225t^2) = \tfrac{1}{4}Kt^2(t - 30)^2$

You can see that $v = 0$ when $t = 0$ and when $t = 30$. These are the only two occasions in the motion when the vehicle is stationary and both correspond to traffic lights, as required.

M2

You can now investigate the displacement of the car. Integrate the expression for the velocity to find the displacement.

$$x = \int K(\tfrac{1}{4}t^4 - 15t^3 + 225t^2)\,dt = K(\tfrac{1}{20}t^5 - \tfrac{15}{4}t^4 + 75t^3) + c'$$

c′ is the constant of integration

The required initial conditions, $x = 0$ when $t = 0$, give $c' = 0$.

So, you have $x = K(\tfrac{1}{20}t^5 - \tfrac{15}{4}t^4 + 75t^3) = \tfrac{1}{20}Kt^3(t^2 - 75t + 1500)$

The other requirement is that $x = 300$ when $t = 30$.

This gives $K = \tfrac{1}{675}$.

The complete model is therefore

$$a = \tfrac{1}{675}t(t - 15)(t - 30)$$
$$v = \tfrac{1}{2700}t^2(t - 30)^2$$
$$x = \tfrac{1}{13\,500}t^3(t^2 - 75t + 1500)$$

These are the graphs of these functions.

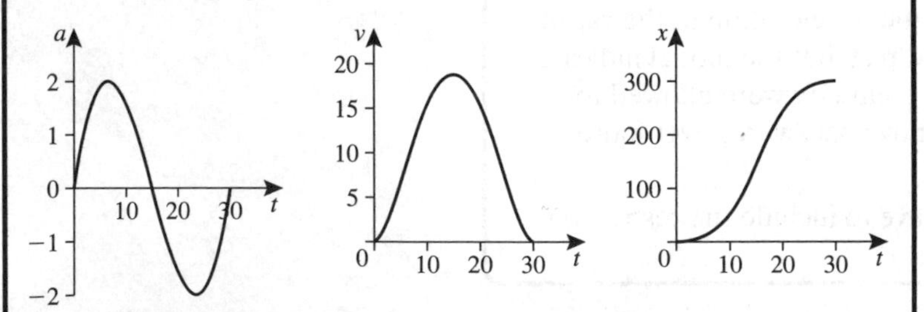

The vehicle reaches its maximum velocity when $a = 0$ at $t = 15$. This gives 18.75 m s^{-1} as the maximum velocity, slightly greater than that suggested by the previous model. This is a reasonable value, but perhaps rather higher than the likely speed limit in an area where most traffic lights are to be found.

30 m.p.h. is roughly 13.4 m s^{-1}

These examples show how it is possible to refine a model to give a more realistic representation of a situation. It would still be necessary to restrict the model to $0 \leqslant t \leqslant 30$, because the second model predicts that, having come to rest, the car immediately starts forward again and continues to accelerate without limit.

Variable forces

Bodies have variable acceleration as a result of their being subjected to variable forces. These can arise in many ways. For example, the motion of a body in a resistive medium such as air or water is opposed by a force which is usually a function of the velocity of the body. Again, the forces involved in magnetic attraction between bodies vary with the distance between them.

In this chapter discussion is restricted to situations where the force varies with time. Some other situations are dealt in Chapter 7.

M2

Example 5

A particle of mass 300 kg, initially at rest, is acted on by a force which is 10 N at the start and increases uniformly to 70 N over 10 seconds. Find the velocity of the particle after 10 seconds.

The force increases uniformly and is therefore a linear function of time. You can write

$$F = pt + q$$

p and q are constants.

When $t = 0$, $F = 10$, and hence $q = 10$.
When $t = 10$, $F = 70$. This gives $10p + 10 = 70$, and hence $p = 6$.
So, you have $F = 6t + 10$.

Call the acceleration a and apply Newton's second law ($F = ma$), so you have

$$6t + 10 = 300a$$

giving $a = \frac{1}{150}(3t + 5)$.

You integrate acceleration to find velocity

$$v = \int \frac{1}{150}\left(3t + 5\right) \mathrm{d}t = \frac{1}{150}\left(\frac{3}{2}t^2 + 5t\right) + c$$

The particle was initially at rest, so $v = 0$ when $t = 0$, giving $c = 0$.
Hence $v = \frac{1}{100}t^2 + \frac{1}{30}t$

At $t = 10$ s, this gives $v = 1\frac{1}{3}\ \mathrm{m\,s^{-1}}$.

Exercise 1A

1 A particle, moving in a straight line, starts from rest at O and has acceleration (in $\mathrm{m\,s^{-2}}$) at time t given by $a = 30 - 6t$.

a) Find its velocity and position at time t.

b) Find its velocity and position after 5 seconds.

c) Find the greatest positive displacement of the particle from O.

d) Find how long the particle takes to return to O.

2 A particle, moving in a straight line, has a velocity given by $v = (6t - 3t^2)\ \text{m s}^{-1}$.

a) Find its change in position from time $t = 1$ to time $t = 3$.

b) Find the distance it travels from time $t = 1$ to time $t = 3$.

3 A particle, P, is moving along a straight wire. At time t seconds, its displacement, x metres from a fixed point, O, on the wire is given by

$$x = t(t^2 - 9)$$

a) Find the time(s) when P is at the fixed point O.

b) Find the time(s) when P is not moving.

c) Find the displacement of P from O when P is stationary.

d) Find the acceleration of P when $t = 5$.

4 The velocity of a particle, travelling along a straight line, is given by $v = 4t + 6$, where the positive direction is to the right. At time $t = 0$, it is 8 m to the left of point A.

a) Find an expression for the position of the particle at time t.

b) Find at what times the particle is at the point A.

c) What is the significance of the negative solution to part b)?

5 A particle is moving along a straight line and its position, measured from the point O, is given by the formula

$$x = t^3 - 2t^2 - t + 2$$

where x is measured in metres and t is measured in seconds.

a) Find the times when the particle is at O.

b) Find the velocities and accelerations at the times when the particle is at O.

6 A particle is moving along a line and has an acceleration given by $a = (2t - 5)\ \text{m s}^{-2}$. When $t = 4$, the particle has a velocity of $2\ \text{m s}^{-1}$ and has a displacement of +8 m.

a) Find an expression for its velocity at time t.

b) Find when the particle is at rest.

c) Find where the particle is when it is at rest.

7 A bird leaves its nest flying along a straight line to an adjacent tree, where it collects some food (without landing). It then returns to its nest along the same line. Its position is modelled by the formula

$$x = 30t - t^2$$

where x measured in metres and t is measured in seconds.

a) How long does the journey take?

b) How far away is the second tree?

c) Criticise the model.

8 A ball is thrown straight up in the air. Its height h m at time t s, measured from the ground, is given by

$$h = 4 + 8t - 5t^2$$

a) Find how long it takes the ball to reach the ground.

b) Find an expression for its velocity, v.

c) Find its velocity when it reaches the ground.

d) Find the maximum height reached by the ball.

e) Explain why the expression $\int_0^t v \; dt$ does not represent the distance travelled by the ball.

9 A safety device is designed to bring a moving body to a stop at a safe distance from the end of a track. The distance, s m, from the end of the track at time t s after the device is activated is modelled by

$$s = 6 + 6e^{-t}$$

a) How far from the end of the track is the body when $t = 0, 1, 2, 3, 4, 5$ seconds?

b) What is the velocity of the body at the times given?

c) What is the acceleration of the body at the times given?

d) Are there any shortcomings with this model?

10 A particle moves so that its acceleration at time t is given by

$$a = -\frac{120}{(t+1)^2}$$

a) Find an expression for the velocity of the particle given that initially it has a velocity of 60 m s^{-1}.

b) Find an expression for the displacement of the particle given that initially it is at the origin.

c) Describe the motion of the particle as t increases.

11 A particle of mass 5 kg is acted on by a variable force $(40t + 10)$ N. The particle is initially at rest. Find its velocity when $t = 5$.

12 A particle of mass 2 kg is initially travelling at 18 m s^{-1}. It is acted upon by a resistive force of magnitude $2t$ N which brings it to rest. Calculate

a) the length of time it takes to stop

b) the distance it travels before coming to rest.

13 A particle of mass 3 kg accelerates from rest under the action of a force of magnitude $\dfrac{6}{t+1}$ N.

Find an expression for its velocity at time t and hence find how much time elapses before its velocity reaches 6 m s^{-1}.

1.2 Motion in two and three dimensions

In two and three dimensions the position (displacement), velocity and acceleration of a particle are vectors, defined relative to some frame of reference. You have already met vectors in two dimensions, which are expressed in terms of the unit vectors **i** in the x-direction and **j** in the y-direction, or as a column vector. For example

$$\mathbf{r} = x\mathbf{i} + y\mathbf{j} = \begin{pmatrix} x \\ y \end{pmatrix}$$

The only change for three dimensions is that you now need a frame of reference with **three perpendicular axes**: the x-, y- and z-axes. The unit vectors in these directions are **i**, **j** and **k** respectively.

If you draw the positive x- and y-axes as usual, there are two possible directions for the positive z-direction: 'out of' the page or 'into' the page. The convention is 'out of' the page, so that if the page is lying on a desk the positive z-axis points straight upwards.

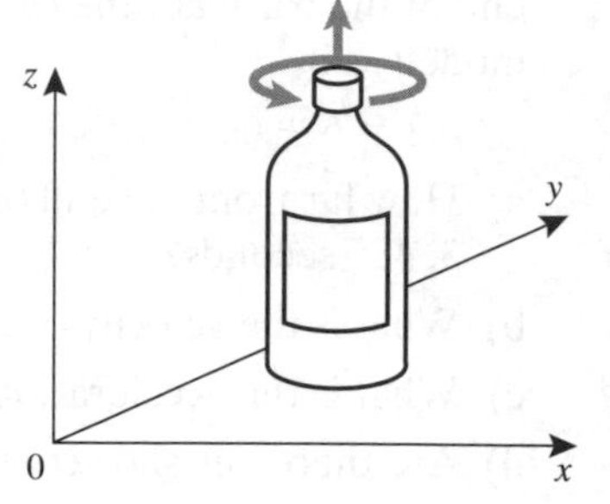

M2

Axes like these form a **right-hand set**, so called because if a right-hand screw (for example, a bottle top) positioned parallel to the z-axis is rotated in the conventional positive (anticlockwise) sense, it will travel in the positive z-direction.

A vector in three dimensions is then expressed in terms of x-, y- and z- components. For example

$$\mathbf{r} = x\mathbf{i} + y\mathbf{j} + z\mathbf{k} = \begin{pmatrix} x \\ y \\ z \end{pmatrix}$$

You can draw 3-D axes in several ways. One common way has been used in the bottle diagram. Another is shown on the right.

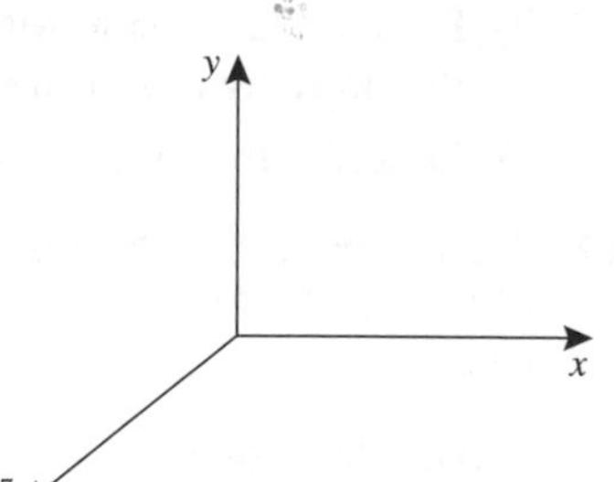

Adding and subtracting vectors in three dimensions, and multiplying by a scalar, are done by separately combining the components. This is exactly the same as in two dimensions.

Finding the magnitude of a vector in three dimensions depends on the use of Pythagoras's theorem, as in two dimensions.

If $\mathbf{r} = a\mathbf{i} + b\mathbf{j} + c\mathbf{k}$, then you have

$$|\mathbf{r}| = \sqrt{a^2 + b^2 + c^2}$$

You can show this using a simple diagram.

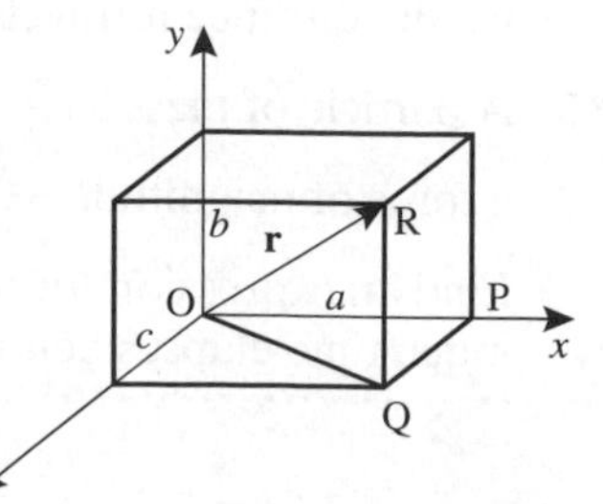

In this diagram, you have a right-angled triangle OPQ, and $OQ^2 = OP^2 + PQ^2 = a^2 + c^2$.

There is also a right-angled triangle OQR, and $OR^2 = OQ^2 + QR^2 = a^2 + c^2 + b^2$

Hence you have $|\mathbf{r}| = OR = \sqrt{a^2 + b^2 + c^2}$

In one-dimensional motion, when you differentiate the displacement of a body with respect to time you find an expression for its velocity. Similarly, when you differentiate its velocity with respect to time you find an expression for its acceleration.

Extending this idea to vectors in general, you can write

$$\mathbf{v} = \frac{\mathrm{d}}{\mathrm{d}t}(\mathbf{r}) = \frac{\mathrm{d}\mathbf{r}}{\mathrm{d}t}$$

$$\mathbf{a} = \frac{\mathrm{d}}{\mathrm{d}t}(\mathbf{v}) = \frac{\mathrm{d}\mathbf{v}}{\mathrm{d}t} = \frac{\mathrm{d}^2\mathbf{r}}{\mathrm{d}t^2}$$

Each component of the velocity is the rate of change of the corresponding displacement component. So, differentiating the displacement vector involves differentiating each component. If the displacement vector is $\mathbf{r} = x\mathbf{i} + y\mathbf{j} + z\mathbf{k}$, you have

$$\mathbf{v} = \frac{\mathrm{d}x}{\mathrm{d}t}\mathbf{i} + \frac{\mathrm{d}y}{\mathrm{d}t}\mathbf{j} + \frac{\mathrm{d}z}{\mathrm{d}t}\mathbf{k}$$

For the acceleration, you have

$$\mathbf{a} = \frac{\mathrm{d}^2x}{\mathrm{d}t^2}\mathbf{i} + \frac{\mathrm{d}^2y}{\mathrm{d}t^2}\mathbf{j} + \frac{\mathrm{d}^2z}{\mathrm{d}t^2}\mathbf{k}$$

There is an alternative notation in which differentiation with respect to t is shown by placing a dot over the variable. You would write

$$\mathbf{v} = \dot{\mathbf{r}} = \dot{x}\mathbf{i} + \dot{y}\mathbf{j} + \dot{z}\mathbf{k}$$

and

$$\mathbf{a} = \ddot{\mathbf{r}} = \ddot{x}\mathbf{i} + \ddot{y}\mathbf{j} + \ddot{z}\mathbf{k}$$

M2

It is necessary that the components are functions of time. To emphasise this the displacement vector is often written as

$$\mathbf{r} = \mathrm{f}(t)\mathbf{i} + \mathrm{g}(t)\mathbf{j} + \mathrm{h}(t)\mathbf{k}.$$

You would then write

$$\mathbf{v} = \mathrm{f}'(t)\mathbf{i} + \mathrm{g}'(t)\mathbf{j} + h'(t)\mathbf{k} \quad \text{and}$$

$$\mathbf{a} = \mathrm{f}''(t)\mathbf{i} + \mathrm{g}''(t)\mathbf{j} + h''(t)\mathbf{k}$$

Example 6

The displacement of a particle at time t s is given by $\mathbf{r} = 4t^2\mathbf{i} + (3t - 5t^3)\mathbf{j}$ metres.
Find expressions for its velocity and acceleration at time t s.

Differentiate displacement to find velocity

$$\mathbf{v} = \frac{\mathrm{d}\mathbf{r}}{\mathrm{d}t} = 8t\mathbf{i} + (3 - 15t^2)\mathbf{j}\ \mathrm{m\,s^{-1}}$$

and a second time to find acceleration

$$\mathbf{a} = \frac{\mathrm{d}\mathbf{v}}{\mathrm{d}t} = 8\mathbf{i} - 30t\mathbf{j}\ \mathrm{m\,s^{-2}}$$

Example 7

The velocity of a particle at time t s is given by
$\mathbf{v} = \cos 2\pi t\mathbf{i} + \sin 2\pi t\mathbf{j} + t^2\mathbf{k}\ \mathrm{m\,s^{-1}}$.
Find its acceleration when $t = 1$.

Differentiate velocity to find acceleration

$$\mathbf{a} = \frac{d\mathbf{v}}{dt} = -2\pi \sin 2\pi t\mathbf{i} + 2\pi \cos 2\pi t\mathbf{j} + 2t\mathbf{k}\ \mathrm{m\,s^{-2}}$$

Substituting $t = 1$ you find $\mathbf{a} = 2\pi\mathbf{j} + 2\mathbf{k}\ \mathrm{m\,s^{-2}}$

Example 8

M2

An aircraft is on a flight path with a position vector relative to a watchtower given by

$$\mathbf{r} = 150t\mathbf{i} + 200t\mathbf{j} + (600 + 50t)\mathbf{k}$$

where the unit vectors **i**, **j** and **k** are measured in the directions east, north and vertically upwards respectively. All distances are measured in metres and all times in seconds.

a) Find expressions for
 i) the aircraft's velocity ii) its acceleration.

b) Find
 i) the position of the aircraft when $t = 0$
 ii) the aircraft's speed and direction.

a) i) Differentiate displacement to find velocity

$$\mathbf{v} = \frac{d\mathbf{r}}{dt} = 150\mathbf{i} + 200\mathbf{j} + 50\mathbf{k}$$

 ii) Differentiate velocity to obtain acceleration

$$\mathbf{a} = \frac{d\mathbf{v}}{dt} = \mathbf{0} \quad \text{(Note that } \mathbf{0} \text{ is the zero vector)}$$

b) i) When $t = 0$, $\mathbf{r} = 600\mathbf{k}$. This means that the aircraft is directly above the watchtower at an altitude of 600 m.

 ii) The velocity of the aircraft is constant.
 The speed is the magnitude of the velocity vector

$$|\mathbf{v}| = \sqrt{150^2 + 200^2 + 50^2} = 255\ \mathrm{m\,s^{-1}}$$

 The direction of travel makes an angle of θ with the **i** (easterly) direction, where $\tan\theta = \dfrac{200}{150}$.

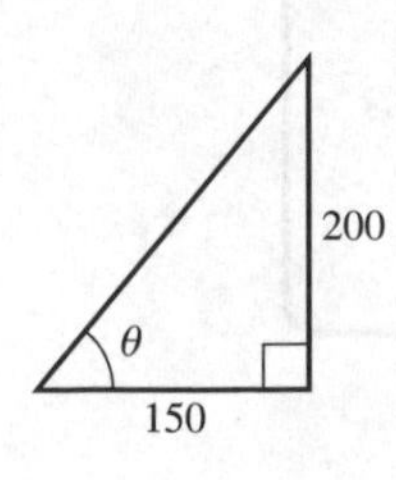

This gives the direction as 53.1° from the **i**-direction. That is a bearing of approximately 037°.
The aircraft is also climbing. The angle of ascent is ϕ, as shown.

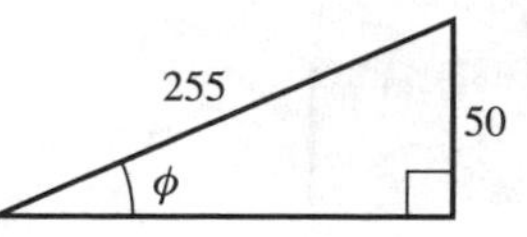

You have $\sin\phi = \dfrac{50}{255}$ and hence $\phi = 11.3°$

Example 9

Two particles, A and B, leave the origin at the same time and move on a plane so that at time t their position vectors are

$\mathbf{r}_A = (8t - 2t^2)\mathbf{i} + (4t + t^2)\mathbf{j}$ and

$\mathbf{r}_B = (4t + t^2)\mathbf{i} + (6t + 1\frac{1}{2}t^2)\mathbf{j}$

Find at what times the particles are moving in the same or opposite directions.

The direction of motion of a particle is determined by its velocity. You differentiate the displacements to find the respective velocities.

$\mathbf{v}_A = (8 - 4t)\mathbf{i} + (4 + 2t)\mathbf{j}$ and

$\mathbf{v}_B = (4+2t)\mathbf{i} + (6 + 3t)\mathbf{j}$

If these velocities have the same or opposite directions then $\mathbf{v}_A = k\mathbf{v}_B$, where k is a constant.
If $k > 0$ they are moving in the same direction.
If $k < 0$ they are moving in opposite directions.

$$k>0 \quad \begin{matrix}\longrightarrow\\ \longrightarrow\end{matrix} \qquad k<0 \quad \begin{matrix}\longrightarrow\\ \longleftarrow\end{matrix}$$

Hence $\quad 8 - 4t = k(4 + 2t)$ and $4 + 2t = k(6 + 3t)$

This gives $\quad k = \dfrac{8 - 4t}{4 + 2t} = \dfrac{4 + 2t}{6 + 3t}$

$$\Rightarrow \quad (8 - 4t)(6 + 3t) = (4 + 2t)^2$$
$$48 - 12t^2 = 16 + 16t + 4t^2$$
$$\Rightarrow \quad t^2 + t - 2 = 0$$

This gives $t = 1$ or $t = -2$.
You can reject the negative root, since it does not satisfy the practical aspects of the problem (and in fact when $t = -2$, $\mathbf{v}_B = \mathbf{0}$).
When $t = 1$, $k = \frac{2}{3}$.
You therefore conclude that when $t = 1$ the particles are moving in the same direction, and there is no time at which they are moving in opposite directions.

In one-dimensional motion, when you integrate the acceleration of a body with respect to time, you find an expression for its velocity. Similarly, when you integrate its velocity with respect to time, you find an expression for its displacement. Extending this idea to vectors in general, you can write

$$\mathbf{v} = \int \mathbf{a}\,\mathrm{d}t \quad \text{and} \quad \mathbf{r} = \int \mathbf{v}\,\mathrm{d}t$$

In a similar way to differentiation, the integration of a vector involves integrating each of its components.

Example 10

The acceleration of a particle at time t is $\mathbf{a} = 4\mathbf{i} + 6t\mathbf{j}$. Find an expression for its velocity.

M2

Integrate acceleration to obtain velocity

$$\mathbf{v} = \int (4\mathbf{i} + 6t\mathbf{j})\,\mathrm{d}t = (4t + c_1)\mathbf{i} + (3t^2 + c_2)\mathbf{j}$$

Remember to add in the integration constants to both components

In practice it is usual to gather together the constant parts

$$\mathbf{v} = 4t\mathbf{i} + 3t^2\mathbf{j} + (c_1\mathbf{i} + c_2\mathbf{j})$$

and hence write

$\mathbf{v} = 4t\mathbf{i} + 3t^2\mathbf{j} + \mathbf{c}$ where the $\mathbf{c}$ is a constant vector.

As with the one-dimensional case, the value of this constant of integration can be found if you know the initial conditions of the problem.

Example 11

The velocity of a particle at time t is given by $\mathbf{v} = 6t^2\,\mathbf{i} + (3t^2 - 8t)\mathbf{j} + 2\mathbf{k}$. Find an expression for its displacement at time t, given that $\mathbf{r} = 5\mathbf{i} - 6\mathbf{j} + \mathbf{k}$ when $t = 0$.

Integrate velocity to find displacement

$$\mathbf{r} = \int (6t^2\mathbf{i} + (3t^2 - 8t)\mathbf{j} + 2\mathbf{k})\,\mathrm{d}t$$
$$= 2t^3\mathbf{i} + (t^3 - 4t^2)\mathbf{j} + 2t\mathbf{k} + \mathbf{c}$$

To find the value of $\mathbf{c}$, use the initial condition $\mathbf{r} = 5\mathbf{i} - 6\mathbf{j} + \mathbf{k}$ when $t = 0$, which gives

$$5\mathbf{i} - 6\mathbf{j} + \mathbf{k} = 0\mathbf{i} + 0\mathbf{j} + 0\mathbf{k} + \mathbf{c}$$

and hence $\quad \mathbf{c} = 5\mathbf{i} - 6\mathbf{j} + \mathbf{k}$

Substituting for $\mathbf{c}$ in the general expression, you find

$$\mathbf{r} = 2t^3\mathbf{i} + (t^3 - 4t^2)\mathbf{j} + 2t\mathbf{k} + 5\mathbf{i} - 6\mathbf{j} + \mathbf{k}$$

or $\quad \mathbf{r} = (2t^3 + 5)\mathbf{i} + (t^3 - 4t^2 - 6)\mathbf{j} + (2t + 1)\mathbf{k}$

Example 12

A particle is moving on a plane with a constant acceleration of $2t\mathbf{j}$. At time $t = 0$ the particle is at the point $\mathbf{i} + 4\mathbf{j}$ and has a velocity of $3\mathbf{i} - 4\mathbf{j}$. Find expressions for

a) the velocity

b) the position of the particle at time t.

a) Integrate acceleration to find velocity

$$\mathbf{v} = \int 2t\mathbf{j}\,\mathrm{d}t = t^2\mathbf{j} + \mathbf{c} \text{ where } \mathbf{c} \text{ is an arbitrary constant.}$$

To find $\mathbf{c}$, you use the initial condition $\mathbf{v} = 3\mathbf{i} - 4\mathbf{j}$ when $t = 0$:

$$3\mathbf{i} - 4\mathbf{j} = 0\mathbf{j} + \mathbf{c} \quad \text{and hence} \quad \mathbf{c} = 3\mathbf{i} - 4\mathbf{j}$$

The velocity is therefore

$$\mathbf{v} = t^2\mathbf{j} + 3\mathbf{i} - 4\mathbf{j} = 3\mathbf{i} + (t^2 - 4)\mathbf{j}$$

b) Integrate velocity to find displacement

$$\mathbf{r} = \int (3\mathbf{i} + (t^2 - 4)\mathbf{j})\,\mathrm{d}t = 3t\mathbf{i} + (\tfrac{1}{3}t^3 - 4t)\mathbf{j} + \mathbf{c}_1$$

where $\mathbf{c}_1$ is a constant.

To find $\mathbf{c}_1$, use the initial condition $\mathbf{r} = \mathbf{i} + 4\mathbf{j}$ when $t = 0$:

$$\mathbf{i} + 4\mathbf{j} = 0\mathbf{i} + 0\mathbf{j} + \mathbf{c}_1 \quad \text{and hence} \quad \mathbf{c}_1 = \mathbf{i} + 4\mathbf{j}$$

The position at time t is therefore

$$\mathbf{r} = 3t\mathbf{i} + (\tfrac{1}{3}t^3 - 4t)\mathbf{j} + \mathbf{i} + 4\mathbf{j}$$
$$= (3t + 1)\mathbf{i} + (\tfrac{1}{3}t^3 - 4t + 4)\mathbf{j}$$

If a body moving in two or three dimensions is acted upon by a variable force then, if the force is a function of time, you can use Newton's second law and integration to find the characteristics of its motion.

Example 13

A particle of mass 4 kg is acted on by a force $(8\mathbf{i} + 12t\mathbf{j})$ N. Initially the particle has velocity $(3\mathbf{i} - 5\mathbf{j})\text{ m s}^{-1}$. Find its velocity after 4 seconds.

By Newton's second law, the equation of motion of the particle is

$$4\mathbf{a} = 8\mathbf{i} + 12t\mathbf{j} \quad \text{which gives} \quad \mathbf{a} = 2\mathbf{i} + 3t\mathbf{j}\text{ m s}^{-2}$$

Integrate acceleration to find velocity

$$\mathbf{v} = \int (2\mathbf{i} + 3t\mathbf{j})\,\mathrm{d}t = 2t\mathbf{i} + \tfrac{3}{2}t^2\mathbf{j} + \mathbf{c}\text{ m s}^{-1}$$

From the initial condition $\mathbf{v} = 3\mathbf{i} - 5\mathbf{j}$ when $t = 0$, you find

$$\mathbf{c} = 3\mathbf{i} - 5\mathbf{j}$$

Hence $\mathbf{v} = 2t\mathbf{i} + \tfrac{3}{2}t^2\mathbf{j} + 3\mathbf{i} - 5\mathbf{j} = (2t + 3)\mathbf{i} + (\tfrac{3}{2}t^2 - 5)\mathbf{j}\text{ m s}^{-1}$

Substituting $t = 4$, you find $\mathbf{v} = 11\mathbf{i} + 19\mathbf{j}\text{ m s}^{-1}$.

M2

Exercise 1B

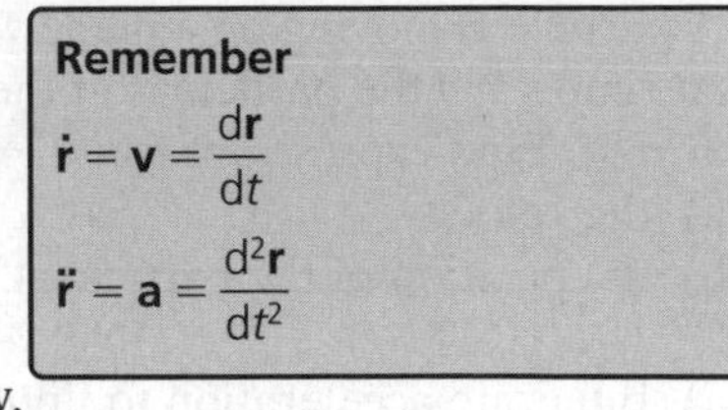

1 For each of the following position vectors, find expressions for the velocity ($\dot{\mathbf{r}}$) and the acceleration ($\ddot{\mathbf{r}}$) of the particle at time t.

a) $\mathbf{r} = 4t\mathbf{i} + (8 + 2t^2)\mathbf{j}$ b) $\mathbf{r} = (t^2 - 4t)\mathbf{i} + (t^3 - 2t^2)\mathbf{j}$

c) $\mathbf{r} = 2\cos t\mathbf{i} + 2\sin t\mathbf{j} + \sqrt{t}\mathbf{k}$ d) $\mathbf{r} = e^t\mathbf{i} + \ln(t + 1)\mathbf{j}$

2 For each of the following velocity vectors, find expressions for the displacement and acceleration at time t, with constants where necessary.

a) $\mathbf{v} = t^3\mathbf{i} + 3t^2\mathbf{j}$ b) $\mathbf{v} = 15\mathbf{i} + (20 - 10t)\mathbf{j}$

c) $\mathbf{v} = (t - t^2)\mathbf{i} + (3t - 5)\mathbf{j}$ d) $\mathbf{v} = 2t(1 - 3t)\mathbf{i} + t^2(3 - 4t)\mathbf{j}$

e) $\mathbf{v} = 4t\mathbf{i} + 4\cos t\mathbf{j} + 2\sin t\mathbf{k}$ f) $\mathbf{v} = -8\sin t\mathbf{i} + 8\cos t\mathbf{j} + 4e^{2t}\mathbf{k}$

3 For each of the following acceleration vectors, find expressions for the velocity and displacement at time t, with constants where necessary.

a) $\mathbf{a} = 6t\mathbf{i} + (2 - 4t)\mathbf{j}$ b) $\mathbf{a} = -10\mathbf{j}$

c) $\mathbf{a} = 2\mathbf{i} + 4t\mathbf{j}$ d) $\mathbf{a} = -5\cos t\mathbf{i} - 5\sin t\mathbf{j} + 6t\mathbf{k}$

4 For the following velocity vectors, find the position vector $\mathbf{r}$ at time t consistent with the given initial condition.

a) $\mathbf{v} = 4t\mathbf{i} + 8t^3\mathbf{j}$, given that $\mathbf{r} = 2\mathbf{i} - \mathbf{j}$ when $t = 0$

b) $\mathbf{v} = 4\sin t\mathbf{i} + 2\mathbf{j}$, given that $\mathbf{r} = 3\mathbf{i} + 2\mathbf{j}$ when $t = 0$

c) $\mathbf{v} = -5\sin t\mathbf{i} - 5\cos t\mathbf{j} + 3\cos 2t\mathbf{k}$, given that $\mathbf{r} = \mathbf{j} - 5\mathbf{k}$ when $t = 0$

5 For the following acceleration vectors, find the velocity vector $\mathbf{v}$ and the position vector $\mathbf{r}$ at time t consistent with the given initial condition.

a) $\mathbf{a} = (3 - 2t)\mathbf{i} + (2t - 6t^3)\mathbf{j}$, given that $\mathbf{v} = 3\mathbf{i}$ and $\mathbf{r} = \mathbf{i} - 2\mathbf{j}$ when $t = 0$

b) $\mathbf{a} = 4\cos 2t\mathbf{i} + 8\sin 2t\mathbf{j}$, given that $\mathbf{v} = 2\mathbf{i} + \mathbf{j}$ and $\mathbf{r} = 4\mathbf{j}$ when $t = 0$

c) $\mathbf{a} = (4t - 3)\mathbf{i} + (6t - 2)\mathbf{k}$, given that $\mathbf{v} = 2\mathbf{i} - 3\mathbf{j}$ and $\mathbf{r} = \mathbf{i} + \mathbf{j} + 2\mathbf{k}$ when $t = 0$

6 A particle moves on a plane such that its position at time t is given by

$$\mathbf{r} = (3t - 2)\mathbf{i} + (4t - 2t^2)\mathbf{j}$$

a) Find expressions for the velocity and acceleration of the particle at time t.

b) Find the initial speed of the particle.

c) At what time(s) is the particle moving parallel to the x-axis?

d) Is the particle ever stationary? Give a reason for your answer.

7 The position of a particle at time t is given by $\mathbf{r} = 2t^3\mathbf{i} + t^2\mathbf{j} - 3t\mathbf{k}$. Find the speed of the particle when $t = 1$.

8 A particle moves on a plane, starting from the point whose position vector is $2\mathbf{i} + 3\mathbf{j}$. Its velocity at time t is given by

$$\mathbf{v} = (4t - 2)\mathbf{i} + 3t^2\mathbf{j}$$

a) Find an expression for the position of the particle at time t.

b) Find the average velocity of the particle over the interval $t = 0$ to $t = 3$.

9 A particle is moving in a plane so that its acceleration is $\mathbf{a} = -2\mathbf{j}$. At time $t = 0$ the particle is at the point whose position vector is $2\mathbf{i} - 3\mathbf{j}$ and has velocity $2\mathbf{i} + 4\mathbf{j}$.

a) Find expressions for its velocity and position at time t.

b) At what time(s) is the particle moving in the **i**-direction?

c) At what time(s) does the particle cross the x-axis?

10 An aircraft is flying at an altitude of 0.8 km, at a speed of 960 km h^{-1}, and on a bearing of 030°. At time $t = 0$ it passes directly over an observer. Taking the observer as the origin and the **i**, **j** and **k**-directions as east, north and upwards respectively:

a) write down the velocity vector, **v**

b) find the position vector, **r**, of the aircraft t hours after passing the observer.

The units for this question are kilometres and hours.

11 A particle of mass 5 kg is acted upon by a force $(20t\mathbf{i} - 15\mathbf{j})$ N. Initially the particle is at the point with position vector $2\mathbf{i} + 3\mathbf{j}$ and is travelling with velocity $(-2\mathbf{i} + 12\mathbf{j})$ m s^{-1}. Find expressions for the velocity and position of the particle at time t, and hence find its velocity and position when $t = 6$.

M2

12 A particle of mass 3 kg is acted upon by a force $(6\mathbf{i} + 3\mathbf{j} + 6(t - 1)\mathbf{k})$ N. Initially the particle is at the origin and is travelling with velocity $(-3\mathbf{j} - 3\mathbf{k})$ m s^{-1} (that is, it is moving perpendicular to the **i**-direction). Show that at one later time it is travelling in the **i**-direction, and find its speed and position at that moment.

13 The position vector of a particle is given by

$$\mathbf{r} = (2 + 3t + 8t^2)\mathbf{i} + (6 + t + 12t^2)\mathbf{j}$$

a) Find an expression for the velocity of the particle at time t.

b) Find an expression for the acceleration of the particle at time t.

An observer placed at the origin sees the particle in the direction given by $\tan\theta = \frac{y}{x}$, where $\mathbf{r} = x\mathbf{i} + y\mathbf{j}$ is the position vector of the particle. The direction in which the particle is moving is given by $\tan\phi = \frac{b}{a}$, where $\mathbf{v} = a\mathbf{i} + b\mathbf{j}$ is the velocity of the particle. The particle will be moving directly away from or directly towards the observer when $\tan\theta = \tan\phi$.

c) Find the time(s) when the particle is moving directly towards or directly away from the observer

d) Find the position of the particle at the time(s) found in c) and identify whether the particle is moving towards or away from the observer.

Summary

You should know how to ...	Check out
1 Find the position, velocity and acceleration of a particle moving with variable acceleration.	**1** A particle starts from rest at time $t = 0$. Its acceleration is $a = 3t^2 + e^{2t}$. Find its velocity and position at time t.
2 Find the position, velocity and acceleration vectors of a particle.	**2** The position of a particle is given by $\mathbf{r} = 3t^2\mathbf{i} + e^{3t}\mathbf{j} - \sin 4t\mathbf{k}$ Find its velocity and acceleration at time t.
3 Apply Newton's laws of motion to situations involving variable acceleration.	**3** A force $(5t^2 + 40 \sin 2t)$ N acts on a body of mass 5 kg. Find the acceleration of the body.

M2

Revision exercise 1

1 A particle moves, so that its acceleration, at time t, is given by

$$\mathbf{a} = -4\cos t\mathbf{i} + 3\sin t\mathbf{j} + \tfrac{1}{2}\mathbf{k}$$

where the unit vectors **i**, **j** and **k** are mutually perpendicular. The initial position of the particle is $4\mathbf{i}$ and its initial velocity is $\frac{1}{2}\mathbf{j}$.

a) Find an expression for the velocity of the particle at time t.
b) Find the position vector of the particle at time t.
c) Find the distance of the particle from the origin when $t = \frac{\pi}{2}$. *(AQA, 2001)*

2 A particle P moves so that at time t seconds its velocity $\mathbf{v}$ m s^{-1} is

$$\mathbf{v} = 2t\mathbf{i} + 4\mathbf{j}, \quad t \geqslant 0.$$

a) At time $t = 0$, the particle is at the point with position vector $2\mathbf{j}$ metres. Find the position vector of P at time t.
b) At times $t = 2$ and $t = 4$, the particle passes through the points C and D respectively. Find the vector $\overrightarrow{CD}$. *(AQA, 2003)*

3 The position vector, **r**, of a particle at time t is given by

$$\mathbf{r} = 4\sin t\mathbf{i} + 4\cos t\mathbf{j} + 6t\mathbf{k}$$

The horizontal unit vectors **i** and **j** are perpendicular and the unit vector **k** is vertical.

a) Find an expression for the velocity of the particle at time t.
b) Find an expression for the acceleration of the particle at time t.
c) Show that the magnitude of the acceleration of the particle is 4.
d) Show that the speed of the particle is constant. *(AQA, 2004)*

4 A particle moves along a straight line. At time t the displacement of the particle from its initial position is x, where

$$x = 4t + 2e^{-t} - 2$$

a) Find the velocity of the particle at time t.
b) Find the acceleration of the particle at time t.
c) Describe what happens to the acceleration of the particle as t increases. *(AQA, 2003)*

5 A particle P moves so that at time t seconds it has position vector $\mathbf{r}$ metres, where

$$\mathbf{r} = (t^4 - 2t^2)\mathbf{i} + (4t^3 - t^4)\mathbf{j}.$$

a) Find an expression for the velocity of P at time t.

The mass of P is 0.25 kg.

b) Find an expression for the momentum of P at time t.
c) Find an expression for the force, $\mathbf{F}$, acting on P at time t.
d) Find the exact value of t when $\mathbf{F}$ acts in the direction of the vector $\mathbf{j}$. *(AQA, 2004)*

6 A particle P moves so that at time t seconds its position vector, $\mathbf{r}$ metres, is

$$\mathbf{r} = (t^3 - 3t)\mathbf{i} + (6t^2 - 12t)\mathbf{j}, \quad t \geqslant 0.$$

a) Use differentiation to find the velocity of P at time t.
b) The mass of P is $\frac{1}{3}$ kg. Express the momentum of P in terms of t.
c) Hence, or otherwise, show that the force, $\mathbf{F}$ newtons, acting on P at time t is

$$2t\mathbf{i} + 4\mathbf{j}.$$

d) Find the magnitude of F when the value of t is 1.5. *(AQA, 2002)*

M2

7 A particle has mass 2000 kg. A single force, $\mathbf{F} = 1000t\mathbf{i} - 5000\mathbf{j}$ N, acts on the particle, at time t seconds. The unit vectors $\mathbf{i}$ and $\mathbf{j}$ are perpendicular. No other forces act on the particle.

a) Find an expression for the acceleration of the particle.
b) At time $t = 0$, the velocity of the particle is $6\mathbf{j}$ m s^{-1}. Show that at time t the velocity, $\mathbf{v}$ m s^{-1}, of the particle is given by

$$\mathbf{v} = \frac{t^2}{4}\mathbf{i} + \left(6 - \frac{5t}{2}\right)\mathbf{j}$$

c) The particle is initially at the origin. Find an expression for the position vector, $\mathbf{r}$ metres, of the particle at time t seconds. *(AQA, 2003)*

8 A cyclist moves from rest along a straight horizontal road. At time t seconds, the displacement of the cyclist from his initial position is s metres.

a) For $0 \leqslant t \leqslant 10$,

$$s = \frac{t^4}{400} - \frac{t^3}{10} + \frac{3t^2}{2}$$

i) Find s when $t = 10$.
ii) Find the velocity of the cyclist when $t = 10$.
iii) Find the acceleration of the cyclist when $t = 10$.

b) For $t \geqslant 10$ the cyclist moves with a constant velocity, so that

$$s = ht - k$$

where h and k are constants. Find the values of h and k. *(AQA, 2004)*

9 A particle P moves so that at time t seconds its position vector, $\mathbf{r}$ metres, is given by

$$r = 6t\mathbf{i} + t^2\mathbf{j}.$$

a) Find the velocity of P at time t.
b) Find an expression for the speed of P at time t.
c) Find the value of t when P moves with speed $6\sqrt{2}$ m s^{-1}. *(AQA, 2004)*

10 At time t, the position vector of a particle Q is

$$\mathbf{r} = (t^2 - 6t + 4)\mathbf{i} + \tfrac{1}{3}t^3\mathbf{j}.$$

a) Find the velocity of Q at time t.

b) Find the value of t when Q is moving parallel to the vector **j**.

c) Find the acceleration of Q and state, with a reason, whether or not this acceleration is constant. *(AQA, 2003)*

11 A boat moves, with constant acceleration, so that at time t its velocity is given by

$$\mathbf{v} = 2(a - t)\mathbf{i} + 4(3 - t)\mathbf{j},$$

where a is a constant and **i** and **j** are unit vectors directed east and north respectively.

a) The boat is heading due north when $t = 2$. Find a.

b) Find:

i) the initial velocity of the boat;

ii) the acceleration of the boat.

M2

c) Find the distance between the initial position of the boat and its position when $t = 2$. *(AQA, 2002)*

12 A particle P moves so that at time t seconds its position vector, **r** metres, is

$$\mathbf{r} = \begin{pmatrix} 2t^2 + 6 \\ 5t \end{pmatrix}, \quad 0 \leqslant t \leqslant 5.$$

a) Find the velocity of P at time t.

b) The force acting on P is $\begin{pmatrix} 2 \\ 0 \end{pmatrix}$ newtons. Find the mass of P.

c) At the instant when $t = 5$, an additional force, $\begin{pmatrix} 0 \\ t \end{pmatrix}$ newtons, begins to act on P.

i) Find the resultant acceleration of P.

ii) Find the velocity of P when $t = 10$. *(AQA, 2002)*

13 A particle P moves so that at time t it has velocity

$$\mathbf{v} = \begin{pmatrix} 2t - 4 \\ 3 \end{pmatrix}, \quad 0 \leqslant t \leqslant 5.$$

a) Given that P has position vector $\begin{pmatrix} 0 \\ 2 \end{pmatrix}$ when $t = 0$, find the position vector of P at time t.

b) The particle P has mass 0.4 units. When $t = 5$, P collides with another particle Q, of mass 0.2 units. Immediately before the collision, Q is moving with velocity $\begin{pmatrix} 12 \\ -3 \end{pmatrix}$.

The particles join together as a result of the collision. Find the velocity of this combined body immediately after the collision. *(AQA, 2001)*

2 Moments

This chapter will show you how to

- ✦ Find the moment of a force about a given point
- ✦ Know the conditions necessary for a rigid body to be in equilibrium
- ✦ Calculate unknown forces acting on a rigid body in equilibrium

Before you start

You should know how to ...	Check in
1 Find components of vectors.	**1** The force **F** has magnitude 8 N and acts at 35° to the **i**-direction, as shown. Express the force **F** in the form $x\mathbf{i} + y\mathbf{j}$.
2 Use the law of friction when a body is in limiting equilibrium.	**2** A block of mass 10 kg rests on a rough horizontal floor. The coefficient of friction is 0.25. A horizontal force of P N is applied and the block is on the point of moving. Find the value of P.

2.1 Turning effects of forces

If a strip of cardboard is pinned to a table top by a single drawing pin so that it is free to move, any horizontal force applied to the strip makes it **rotate**. The only exception is if the force acts along a line which passes **through** the drawing pin. The strength of this turning effect depends on

- ✦ the **magnitude** of the force applied,
- ✦ **where** it is applied, and
- ✦ in **which direction** it is applied.

To explore the way that a force and its point of application are related to the strength of the turning effect you can use the following experiment.

Rod and pivot experiment

You could use a piece of wood.

You need a rigid uniform rod at least 1 m long. Fix a pivot to the rod so that it is free to rotate in a vertical plane about its centre. The rod should stay in a horizontal position when placed there.

Hang two equal masses at equal distances on either side of the pivot. The rod should remain horizontal because the weights acting on each side have **equal** and **opposite** turning effects.

Now increase one of the masses so that the rod no longer balances. Move this larger mass towards the pivot to restore the balance. Record its new distance from the pivot.

Repeat this so that you have data for eight different masses.

Make sure you change the mass on the same side each time.

When the author tried this experiment, the original masses were 40 g and were initially placed at a distance of 48 cm either side of the central pivot.

The initial balancing forces were

$0.04g$ N at 0.48 m balancing $0.04g$ N at 0.48 m.

Use Newton's second law: $F = ma$

One mass was altered and moved, so to balance the rod, giving

$0.04g$ N at 0.48 m balancing mg N at x m.

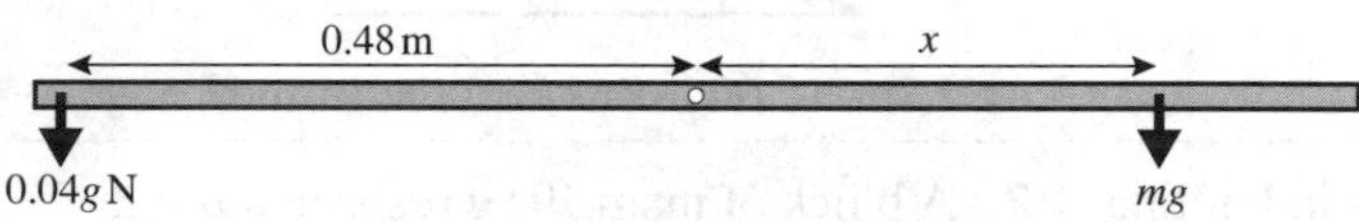

The table and graph show the results.

Distance (x)	Mass (kg)	Force = mg (N)	Force × Distance (N m)	
x	m	F	Fx	$\frac{1}{x}$
0.48	0.04	0.392	0.19	2.08
0.32	0.06	0.588	0.19	3.13
0.24	0.08	0.784	0.19	4.17
0.19	0.10	0.98	0.19	5.26
0.16	0.12	1.176	0.19	6.25
0.14	0.14	1.372	0.19	7.14
0.12	0.16	1.568	0.19	8.33
0.11	0.18	1.764	0.19	9.09

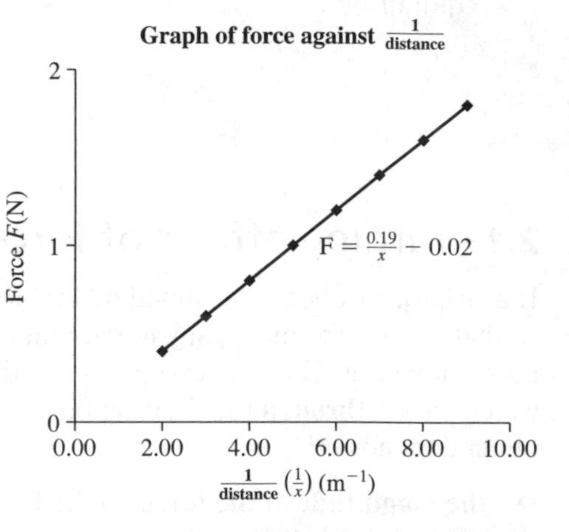

The data show pairs of force and distance that each generate the same turning effect.

It is clear from these data that the product **force × distance = constant**.

That is $Fx = k$

or $F = \frac{k}{x}$

This indicates that plotting a graph of F against $\frac{1}{x}$ should give a straight line through the origin. The graph of the data confirms this to a good level of accuracy.

If you perform the experiment for yourself, using the spreadsheet MOMENTS1, available on the OUP website, you can carry out the above analysis of your data. Just type in: http://www.oup.co.uk/secondary/mechanics

M2

In the above experiment you found that force × distance remained constant. The constant property of the data pairs was the turning effect. You can write this as the following definition:

Moment of a force

When a force **F** acts in a plane at a perpendicular distance d from an axis that is perpendicular to the plane, the turning effect of the force about the axis has magnitude $|\mathbf{F}| \times d$. This turning effect is called the **moment of the force** or the **torque**.

As $|\mathbf{F}|$ is in newtons and d is in metres, the moment of a force is measured in newton metres (N m).

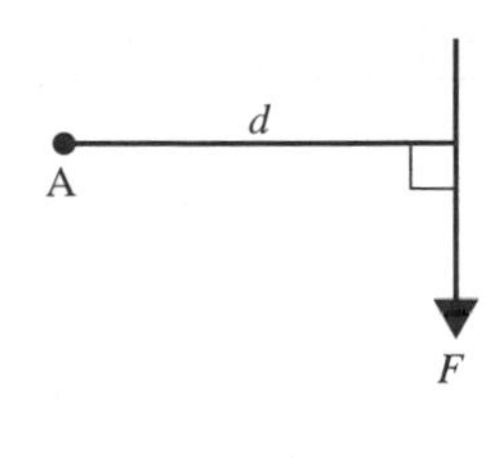

Although you should strictly always refer to the moment of a force about an axis, all the situations dealt with in this specification involve forces acting in a single plane. You can therefore talk of the moment of a force about a point.

You also need to be clear about the direction of the turning effect. In the two situations illustrated, the **magnitude** of the turning effect is the same. However, in case a), the rotation is anticlockwise, whereas in case b) the rotation is clockwise. It is usual to regard anticlockwise as the **positive** rotational direction, so the moment in case a) is $+Fd$ and in case b) is $-Fd$.

In most situations there will be two or more forces acting, each with its own turning effect about the chosen point. These combine to given an overall turning effect. That is:

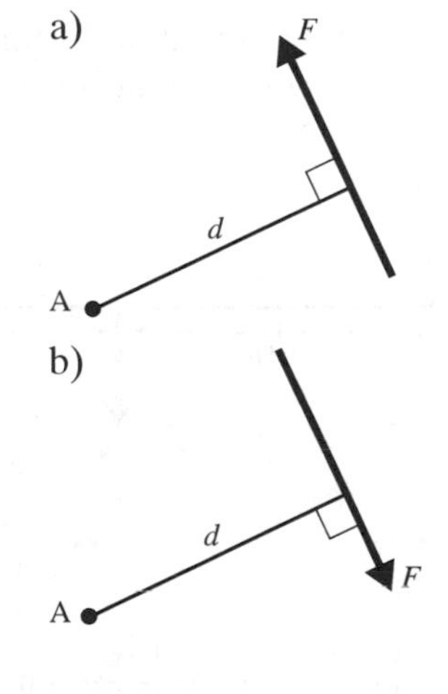

When several forces act in the same plane, their **total turning effect** about a given point is the **sum of the moments of the individual forces**.

M2

Example 1

Forces of 10 N, 15 N and 18 N act as shown on a rectangular lamina ABCD with AB = 6 m and BC = 4 m. Find the total moment of the forces about A.

Note The term *lamina* means an idealised, infinitely thin, plane object. The plural is *laminae* or *laminas*.

The 18 N force acts along a line 4 m from A and turns anticlockwise. Therefore, its moment is

$$+18 \times 4 = +72 \text{ N m}$$

The 10 N force acts along a line 6 m from A and turns clockwise. Therefore, its moment is

$$-10 \times 6 = -60 \text{ N m}$$

The 15 N force acts directly through A and so has no turning effect. Its moment is 0 N m.

The total moment of the forces is, therefore, $72 - 60 + 0 = 12$ N m

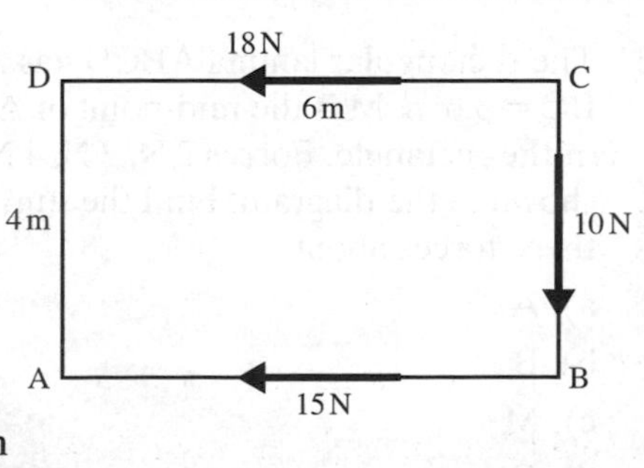

Exercise 2A

1 Find the moment of each of the following forces about the point A, indicating whether the moment is positive or negative.

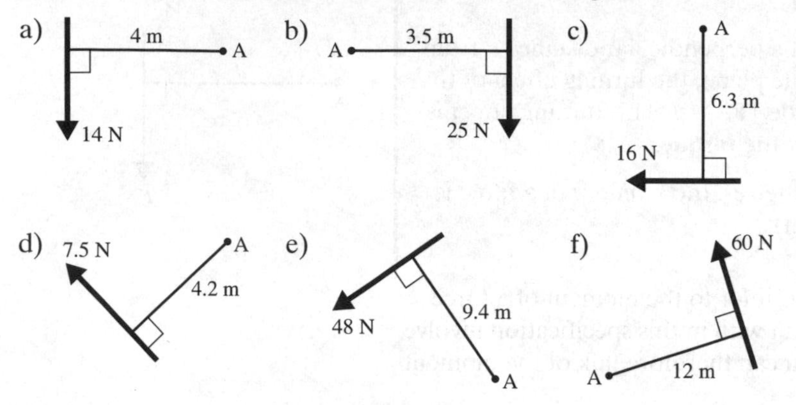

2 Find the total moment of the forces shown in each of the following diagrams about i) A and ii) B.

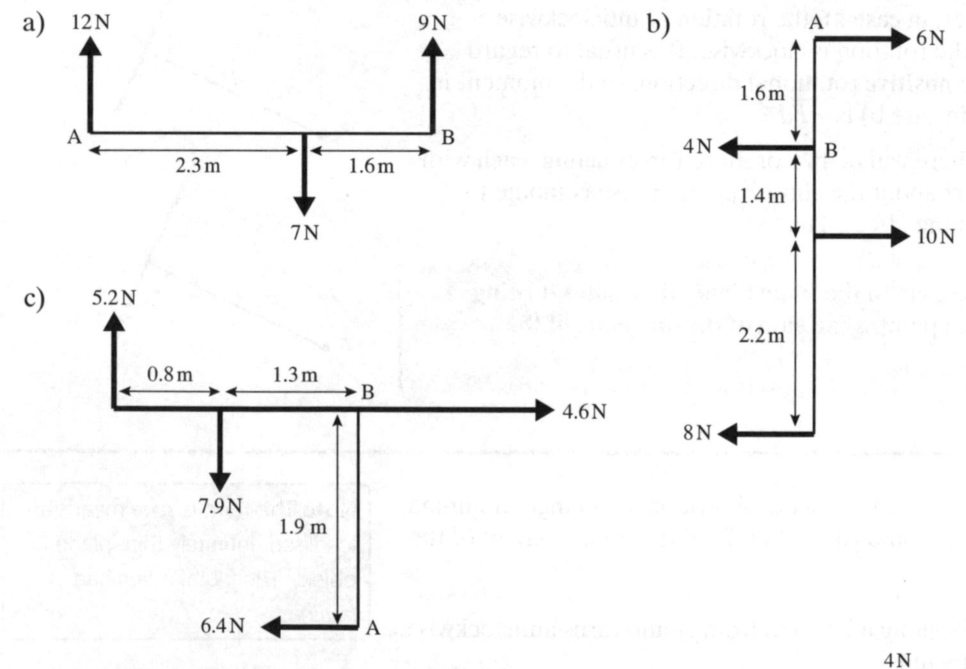

3 The rectangular lamina ABCD has AB = 4.8 m and BC = 3.6 m. M is the mid-point of AB and O is the centre of the rectangle. Forces 2 N, 3 N, 4 N and 6 N act as shown in the diagram. Find the sum of the moments of these forces about

a) A

b) B

c) M

d) O

4 The diagram shows an aerial view of a revolving door. Four people are exerting forces of 40 N, 60 N, 80 N and 90 N as shown. Find the distance x if the total moment of the forces about 0 is

a) 12 N m

b) −8 N m

c) 0 N m

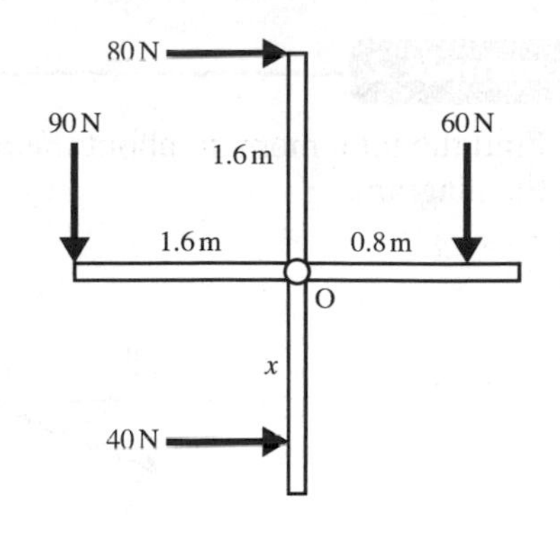

Forces at an angle

The diagram shows a rod AB of length a. A force of magnitude F is applied at B at an angle θ to the rod.

There are two ways of finding the moment of F about A.

Method 1
The perpendicular distance from A to the line of action of F is AC, so

Moment of F about A $= F \times \text{AC}$
$= Fa \sin\theta$

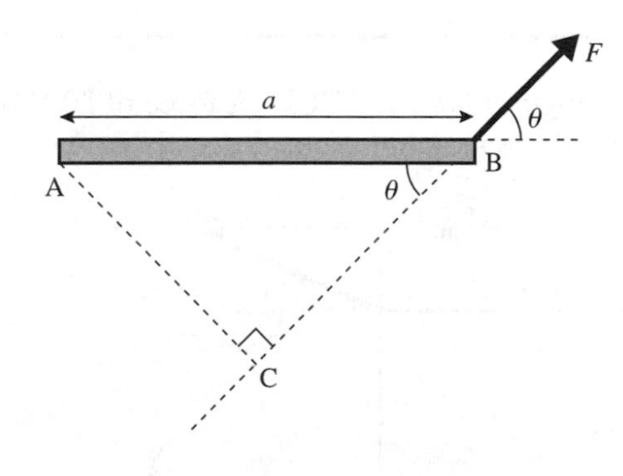

Method 2
The force F can be resolved into two components:

$F\cos\theta$ in the direction AB
$F\sin\theta$ perpendicular to AB

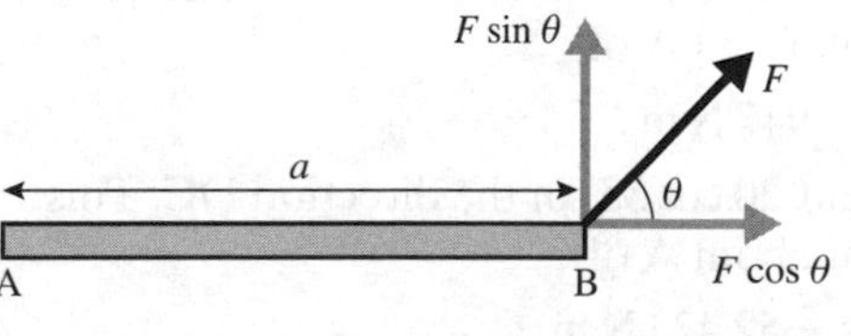

The first component is along a line through A and so has no turning effect about A. The moment of F is produced entirely by the second component. So,

$$\text{Moment of } F \text{ about A} = F\sin\theta \times a = Fa\sin\theta$$

You should be able to use both methods, so that you can choose the better one depending on the circumstances.

M2

Example 2

Find the total moment about the point A of the forces shown in the diagram.

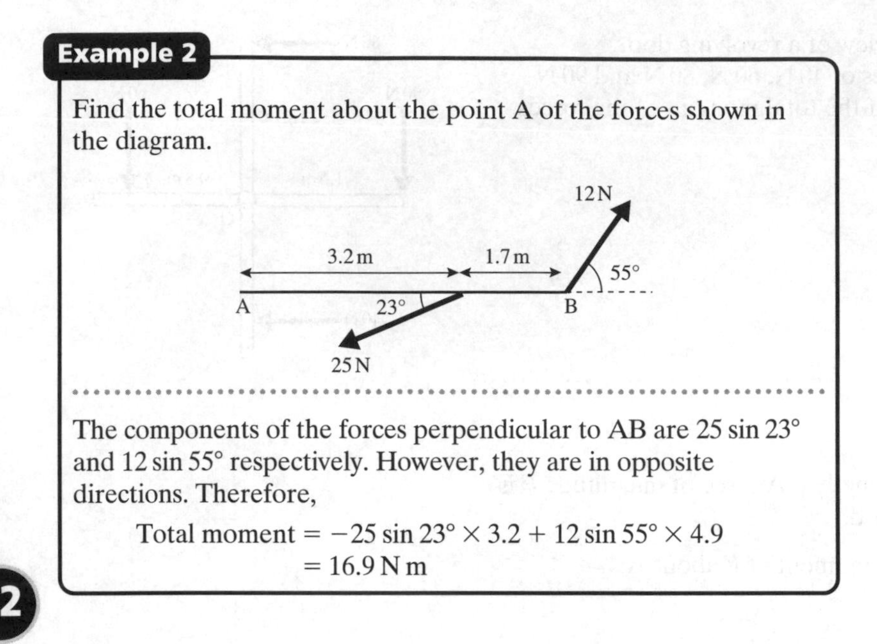

The components of the forces perpendicular to AB are 25 sin 23° and 12 sin 55° respectively. However, they are in opposite directions. Therefore,

$$\text{Total moment} = -25 \sin 23° \times 3.2 + 12 \sin 55° \times 4.9$$
$$= 16.9 \text{ N m}$$

Example 3

The diagram shows a rectangular lamina ABCD. A force of 20 N is applied at C, as shown. Find the moment of this force about A.

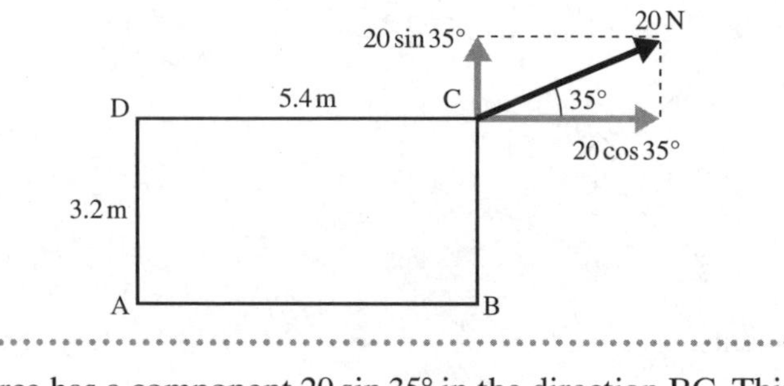

The force has a component 20 sin 35° in the direction BC. This component has a moment about A of

$$20 \sin 35° \times 5.4 = 61.947 \text{ N m}$$

The force has a component 20 cos 35° in the direction DC. This component has a moment about A of

$$-20 \cos 35° \times 3.2 = -52.423 \text{ N m}$$

Therefore, the total moment about A is

$$61.947 - 52.423 = 9.52 \text{ N m}$$

Forces in vector component form

Sometimes, forces may be given in the vector component form $a\mathbf{i} + b\mathbf{j}$, and points given as either cartesian coordinates (x, y) or position vectors $x\mathbf{i} + y\mathbf{j}$.

This form makes it easier to find the moments of the component forces.

Example 4

Find the moment about the point P, that has position vector $(\mathbf{i} + 3\mathbf{j})$ m, of the force $(5\mathbf{i} + 2\mathbf{j})$ N acting at the point Q, that has position vector $(4\mathbf{i} + 5\mathbf{j})$ m.

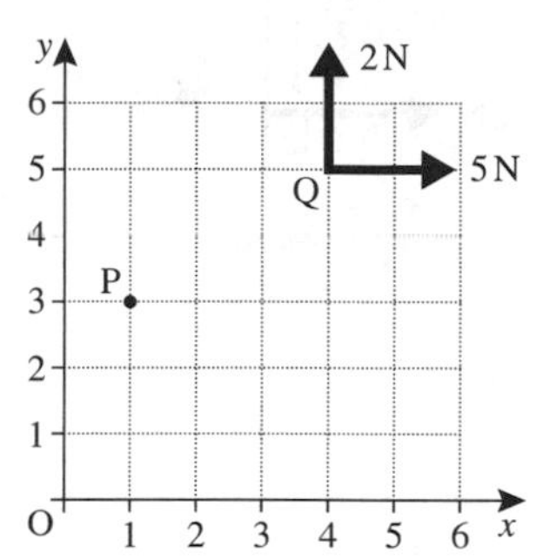

From the diagram, you can see that:

The 5 N component has a clockwise moment about P of $-5 \times 2 = -10$ N m

The 2 N component has an anticlockwise moment about P of $2 \times 3 = 6$ N m

Therefore, the total moment of the force about P is

$-10 + 6 = -4$ N m

Exercise 2B

1 Find the moment of each of the following forces about A, indicating whether it is a positive or negative moment.

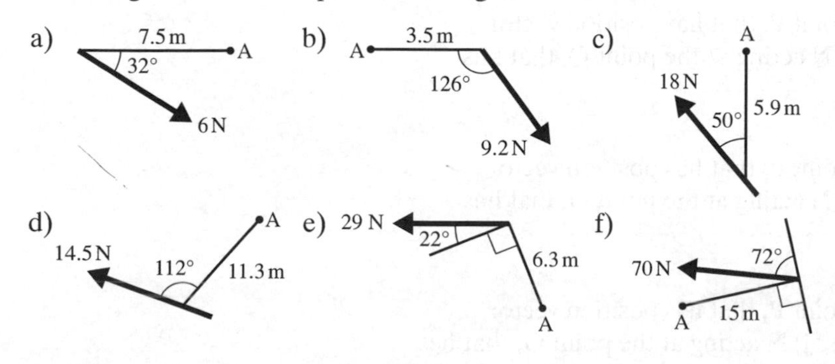

2 Find the total moment of the forces shown in each of the following diagrams about i) A and ii) B.

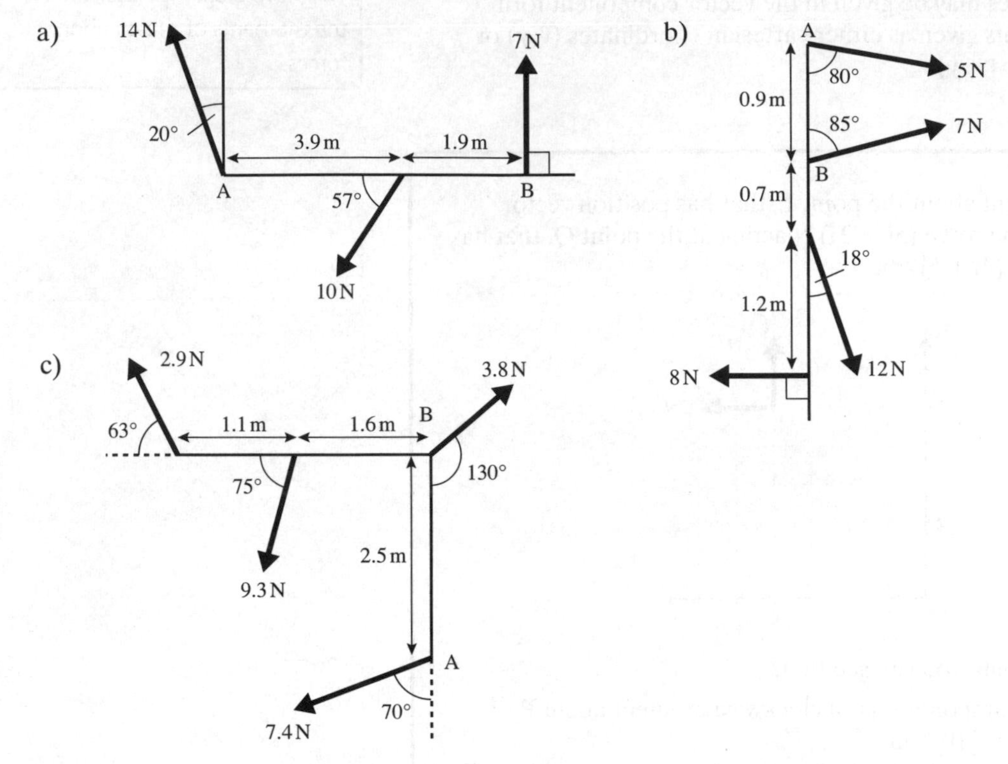

3 A force of 12 N acts along the diagonal AC of a rectangular lamina ABCD, as shown. Find the moment of the force about B.

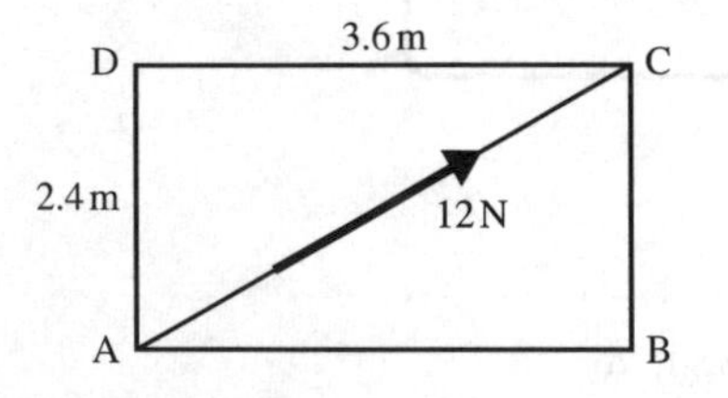

4 Find the moment about the point P, that has position vector $(2\mathbf{i} + \mathbf{j})$ m, of a force $(4\mathbf{i} + 6\mathbf{j})$ N acting at the point Q, that has position vector $(4\mathbf{i} + 7\mathbf{j})$ m.

5 Find the moment about the point P, that has position vector $(\mathbf{i} - 3\mathbf{j})$ m, of a force $(2\mathbf{i} - 3\mathbf{j})$ N acting at the point Q, that has position vector $(2\mathbf{i} + 4\mathbf{j})$ m.

6 Find the moment about the point P, that has position vector $(-5\mathbf{i} + \mathbf{j})$ m, of a force $(-3\mathbf{i} + 5\mathbf{j})$ N acting at the point Q, that has position vector $(-3\mathbf{i} - 5\mathbf{j})$ m.

2.2 Parallel forces

Principle of moments

You know that any system of forces is equivalent to a single force, called the **resultant**. The combined **translational** effect of the forces is the same, both in magnitude and direction, as that of their resultant.

It is reasonable to assume that the combined **rotational** effect of a system of forces about any point is also the same as that of their resultant. This is called the **principle of moments**. That is:

You can turn to the appendix to this book for a proof of this assumption – see pages 169–170

The **total moment of a system of forces** about any point is equal to the **moment of the resultant force of the system** about that point.

Resultant of parallel forces

When forces act along **parallel lines** their resultant is just the sum of the forces.
There are three possible situations.

a) The sum of the forces is non-zero. The system of forces is equivalent to a single resultant force, and you can use the principle of moments to find the line along which it acts.
b) The sum of the forces is zero and the total moment of the forces about any point is also zero. The system of forces is then in equilibrium.
c) The sum of the forces is zero but the total moment of the forces is non-zero. The simplest example of this is when there are two equal forces acting in parallel but opposite directions. Such a system has a pure turning effect and is called a **couple**.

Depending on their direction, some of the forces may be negative. If two parallel forces act in the same direction they are called **like** forces, whereas when they act in opposite directions they are **unlike** forces.

You will not be required to deal with couples in this specification.

M2

Example 5

The diagram shows a light rod AB of length 3 m. C is the point on the rod such that AC = 2 m. Forces of 6 N, 7 N and 8 N act at A, B and C, as shown. Find the resultant, R, of this system of forces, and determine the distance, x m, of its line of action from A.

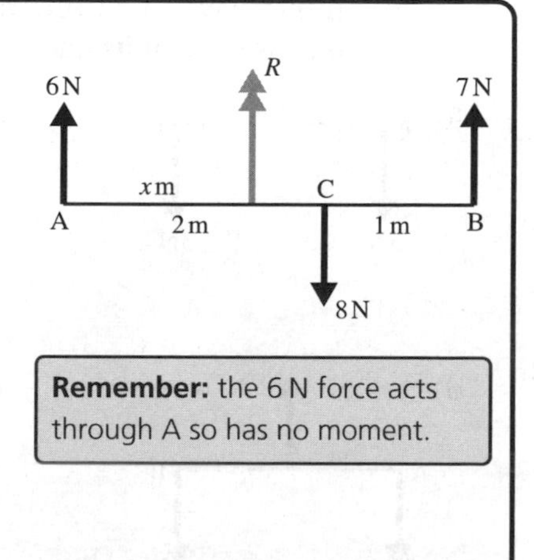

Resolving perpendicular to AB, you have

$$R = 6 + 7 - 8 = 5\text{ N}$$

The total moment of the system of forces about A is

$$7 \times 3 - 8 \times 2 = 5\text{ N m}$$

By the principle of moments this equals the moment of the resultant about A, giving

$$5 \times x = 5 \quad \text{and so} \quad x = 1\text{ m}$$

Remember: the 6 N force acts through A so has no moment.

Example 6

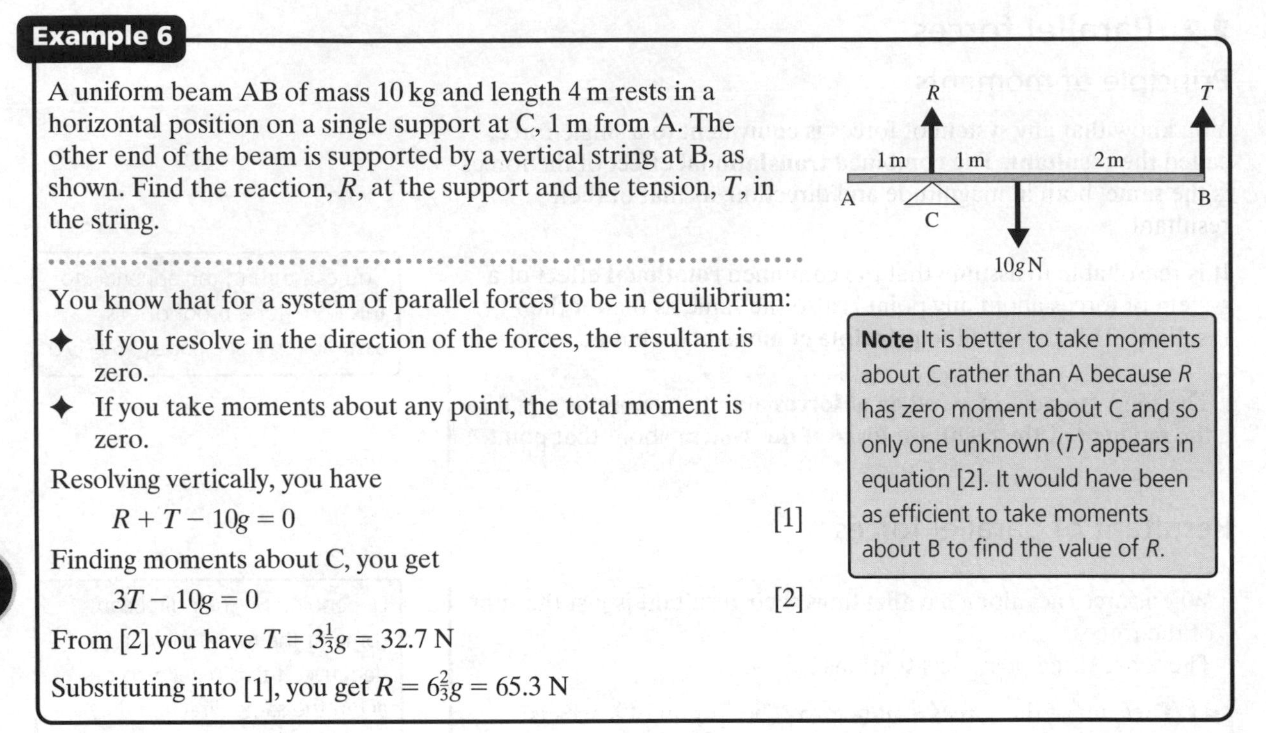

A uniform beam AB of mass 10 kg and length 4 m rests in a horizontal position on a single support at C, 1 m from A. The other end of the beam is supported by a vertical string at B, as shown. Find the reaction, R, at the support and the tension, T, in the string.

You know that for a system of parallel forces to be in equilibrium:

- If you resolve in the direction of the forces, the resultant is zero.
- If you take moments about any point, the total moment is zero.

Resolving vertically, you have

$$R + T - 10g = 0 \qquad [1]$$

Finding moments about C, you get

$$3T - 10g = 0 \qquad [2]$$

From [2] you have $T = 3\frac{1}{3}g = 32.7$ N

Substituting into [1], you get $R = 6\frac{2}{3}g = 65.3$ N

Note It is better to take moments about C rather than A because R has zero moment about C and so only one unknown (T) appears in equation [2]. It would have been as efficient to take moments about B to find the value of R.

M2

Exercise 2C

1 The following diagrams show a light rod AB, of length 4 m, acted upon by parallel forces perpendicular to it.

i) By resolving perpendicular to the rod, and finding the total moment of the system of forces about some point, decide whether the forces have a non-zero resultant, are in equilibrium or neither (i.e. they form a couple).

ii) If they have a non-zero resultant, find its magnitude and the distance of its line of action from A and B.

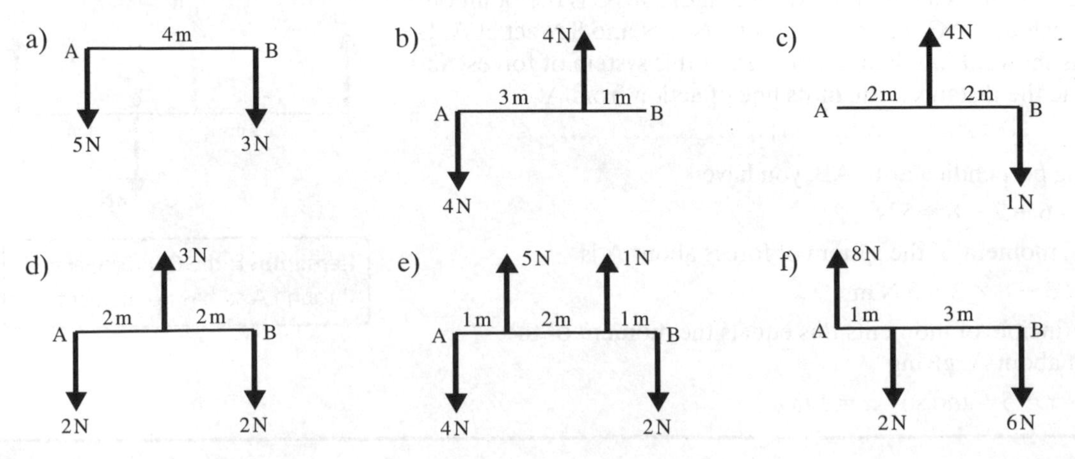

2 The diagram below shows forces acting on a rod AB of length 5 m.

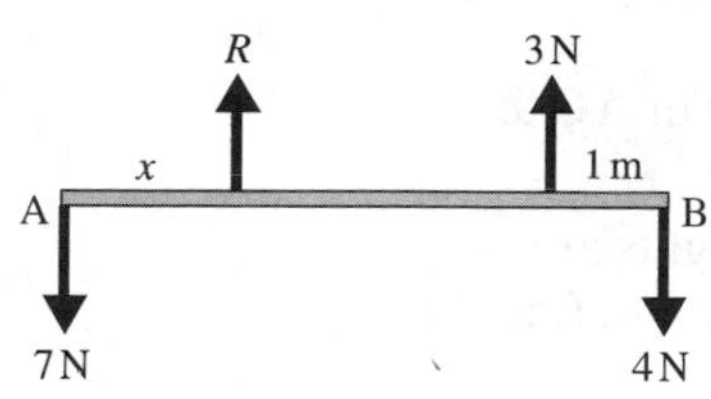

Find the values of R and x when the forces

a) are in equilibrium

b) have a resultant of magnitude 2 N acting downwards through the mid-point of AB.

3 A light rod AB of length 2 m is suspended by two vertical strings at A and B.

a) An object of weight 200 N is suspended from the rod at C, where AC = 0.5 m. Calculate the tensions in the strings.

b) The strings have a breaking strain of 180 N. The object is gradually moved along the rod towards A. How close can it get to A before the string breaks?

4 A uniform beam AB of length 4 m and mass 50 kg rests on supports at A and B. Objects of mass 20 kg and 40 kg are hung on the beam at C and D respectively, where AC = 1.4 m and AD = 3.2 m. Find the reactions at the supports.

5 A uniform rod AB of length $12a$ and weight $4W$ is suspended by strings attached at C and D, where AC = $3a$ and BD = $4a$. The breaking strain of the string at C is $3W$ and that at D is $3.8W$. An object of weight W is attached to the beam at a distance x from A. Find the range of values of x if neither string is to break.

6 A non-uniform beam AB, of length a, has weight W which acts through its centre of gravity a distance b from A. It is placed symmetrically on two supports a distance c apart. Find, in terms of W, a, b and c, the reactions at the supports.

7 A uniform plank is 12 m long and has mass 100 kg. It is placed on horizontal ground at the edge of a cliff, with 4 m of the plank projecting over the edge.

a) How far out from the cliff can a man of mass 75 kg safely walk?

b) The man wishes to walk to the end of the plank. What is the minimum mass he should place on the other end of the plank so he may safely do this?

8 A heavy beam AB rests on two supports at points C and D, where CD = a. An object of weight W rests on the beam. If the object is moved a distance b in the direction DC, show that, provided equilibrium is maintained, the reaction at C will be increased by $\dfrac{Wb}{a}$.

9 A rectangular lamina ABCD is free to rotate in a vertical plane about its centre O. Weights of W, $5W$, $2W$ and $3W$ are attached at A, B, C and D respectively.

a) If the system is in equilibrium, find the inclination of AC to the horizontal.

b) Show that, by changing the order in which the weights are attached, a second equilibrium position is possible and find the inclination of the diagonal in this case.

2.3 Equilibrium of non-parallel forces

As with parallel forces, a system of non-parallel forces corresponds to a single resultant force to give a translational effect. If this resultant is zero there may still be a rotational effect (a couple). To be in equilibrium, both the translational and rotational effects must be zero. This is the case if the following two conditions apply.

M2

- The resultant force **in each of two non-parallel directions** is zero.
- The total moment about any chosen point is zero.

It is necessary to resolve in two directions because the system of forces may have a non-zero resultant perpendicular to the first direction chosen.

These are the most commonly used **conditions for equilibrium.**

It follows that for a system in equilibrium you can find three equations, by resolving in two different directions and taking moments about a point.

Of course you could go on and resolve in other directions or take moments about other points, but the equations would contain no extra information. They would just be combinations of the original three equations.

Example 7

A ladder AB of mass 20 kg rests on smooth horizontal ground and leans against a smooth vertical wall. The inclination of the ladder to the horizontal is 60°. The ladder is kept in position by a horizontal force P N applied to the bottom of the ladder. Find the value of P and the reactions at the wall and the ground.

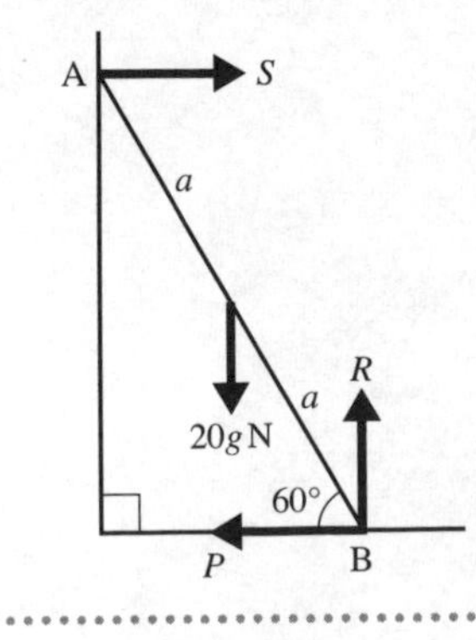

Suppose the ladder is of length $2a$. The reactions R and S at the ground and the wall are normal because the contacts are smooth. The system is in equilibrium.

Resolving vertically, you get

$$R - 20g = 0 \quad \text{and so} \quad R = 196\text{ N}$$

Resolving horizontally, you get

$$P - S = 0 \qquad [1]$$

Taking moments about B, you have

$$20g\cos 60° \times a - S\sin 60° \times 2a = 0 \qquad [2]$$

From [2] $\quad 10g = S\sqrt{3} \quad$ and so $\quad S = 56.6\text{ N}$

From [1] $\quad P = 56.6\text{ N}$

Note You should always be careful in your choice of the directions for resolving and of the point about which you take moments. For instance, you could resolve parallel and perpendicular to the ladder and take moments about the middle of the ladder. The resulting equations would be

$$S\cos 60° + 20g\sin 60° - R\sin 60° - P\cos 60° = 0$$

$$S\sin 60° + R\cos 60° - 20g\cos 60° - P\sin 60° = 0$$

$$Ra\cos 60° - Sa\sin 60° - Pa\sin 60° = 0$$

These equations still give the correct values, but the algebra involved is more tedious.

Alternative conditions for equilibrium

You know that you can find three independent equations by equating to zero the total components in two chosen directions and the total moment about one chosen point. You can achieve a similar result in two other ways, which can occasionally prove more convenient.

- ✦ **Method 2** Take moments about two points and resolve in one direction.
- ✦ **Method 3** Take moments about three points.

In Method 2, if you take moments about P and Q, say, you must not resolve perpendicular to PQ. Otherwise a system with a single force along PQ would appear to be in equilibrium.

In Method 3, the three points must not be in a straight line (for a similar reason).

M2

Example 8

A uniform rod AB, of length 2 m and mass 5 kg, rests with A on smooth horizontal ground and B on a rough peg 1 m above the ground. Find the reaction at A and the normal reaction and friction forces at B.

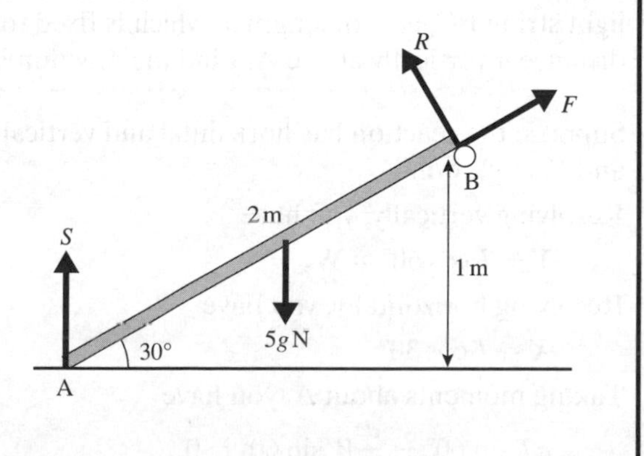

From the dimensions given, the angle at A is 30°.

Taking moments about B, you have

$$5g\cos 30° \times 1 - S\cos 30° \times 2 = 0$$

giving $S = 2.5g = 24.5\text{ N}$

Taking moments about A, you have

$$R \times 2 - 5g\cos 30° \times 1 = 0$$

giving $R = 2.5g\cos 30° = 21.2\text{ N}$

Resolving parallel to AB, you get

$$F + S\sin 30° - 5g\sin 30° = 0$$

giving $F = 2.5g\sin 30° = 12.25\text{ N}$

Example 9

A fixed smooth cylinder, radius a and centre O, rests on a smooth horizontal surface with its axis horizontal. A rod AB of weight W rests with A on the horizontal surface and B on the cylinder such that AB is inclined at 60° to the horizontal and is a tangent to the cylinder. The rod is held in place by a light string, AP, attached to the cylinder at P so that APO is a straight line. Find the tension in the string and the reactions at A and B.

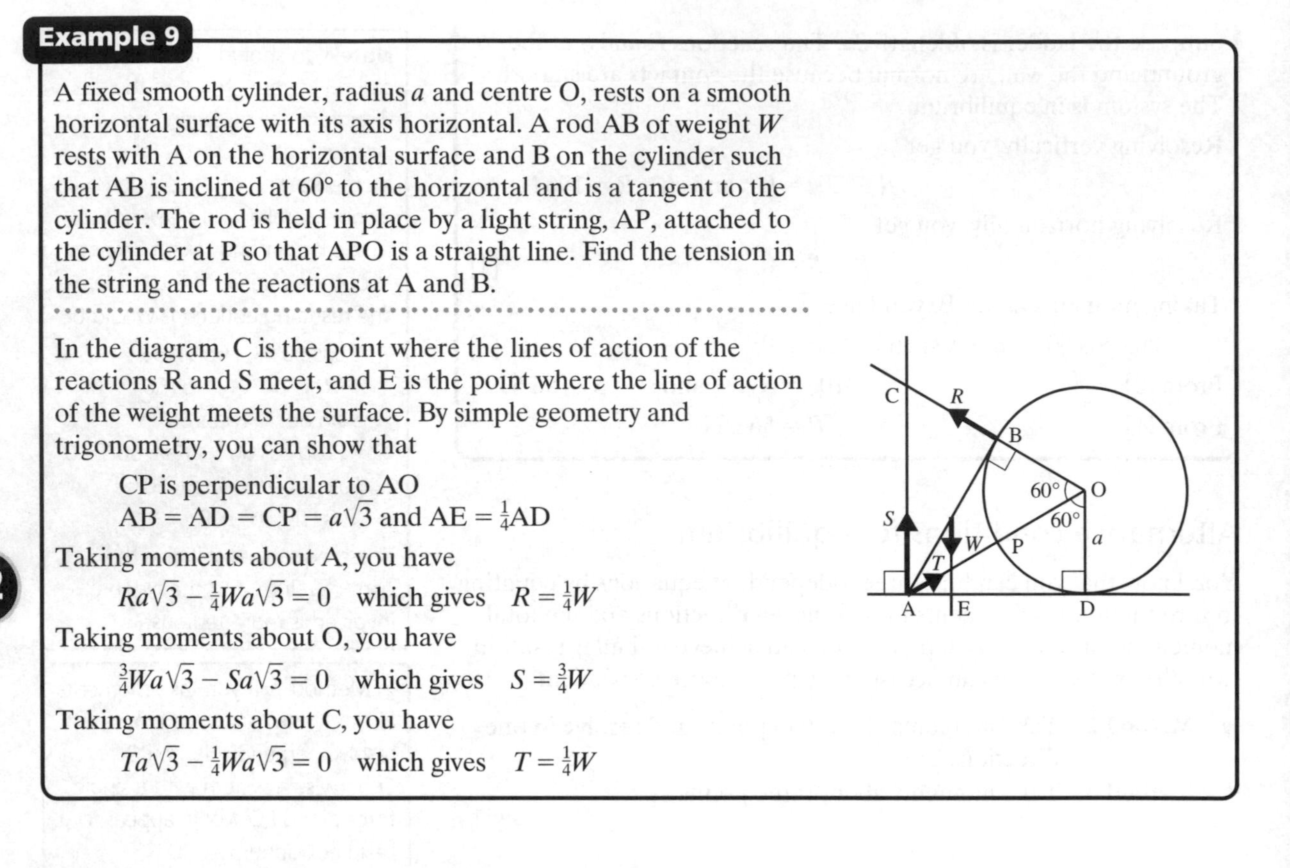

In the diagram, C is the point where the lines of action of the reactions R and S meet, and E is the point where the line of action of the weight meets the surface. By simple geometry and trigonometry, you can show that

CP is perpendicular to AO

$\text{AB} = \text{AD} = \text{CP} = a\sqrt{3}$ and $\text{AE} = \frac{1}{4}\text{AD}$

M2

Taking moments about A, you have

$Ra\sqrt{3} - \frac{1}{4}Wa\sqrt{3} = 0$ which gives $R = \frac{1}{4}W$

Taking moments about O, you have

$\frac{3}{4}Wa\sqrt{3} - Sa\sqrt{3} = 0$ which gives $S = \frac{3}{4}W$

Taking moments about C, you have

$Ta\sqrt{3} - \frac{1}{4}Wa\sqrt{3} = 0$ which gives $T = \frac{1}{4}W$

Reactions in hinges and joints

Some problems require you to find the **reaction** in a hinge or a joint. This is usually best achieved by finding the horizontal and vertical components of the reaction.

Example 10

A uniform rod AB of length a and weight W is hinged to a vertical wall at A and is held at an angle of 30° above the horizontal by a light string BC, also of length a, which is fixed to the wall at C, a distance a vertically above A. Find the reaction in the hinge at A.

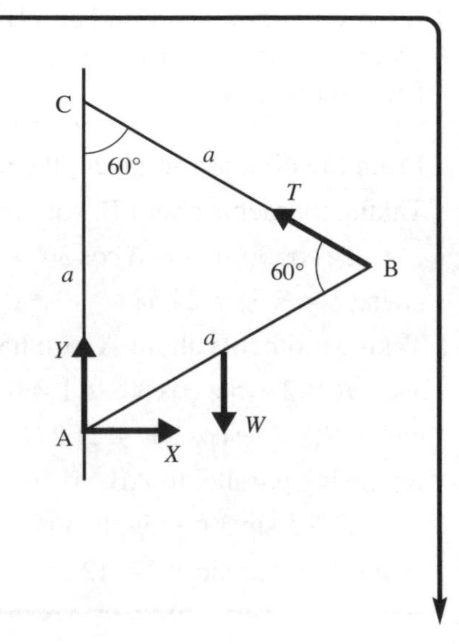

Suppose the reaction has horizontal and vertical components, X and Y, as shown.

Resolving vertically, you have

$$Y + T\cos 60° = \text{W} \quad [1]$$

Resolving horizontally, you have

$$X = T\cos 30° \quad [2]$$

Taking moments about A, you have

$$aT\sin 60° - \frac{a}{2}W\sin 60° = 0 \quad [3]$$

From [3], $T = \frac{1}{2}W$

From [2], $X = \frac{\sqrt{3}}{2}T = \frac{\sqrt{3}}{4}W$

From [1], $Y = W - \frac{1}{2}W.\frac{1}{2} = \frac{3}{4}W$

You can now combine X and Y to find the magnitude of the reaction R

$$R = \sqrt{X^2 + Y^2} = \tfrac{1}{2}W\sqrt{3}$$

The reaction makes an angle θ to the horizontal, where

$$\tan\theta = \frac{Y}{X} = \sqrt{3} \quad \text{and hence} \quad \theta = 60°$$

Three forces in equilibrium

You have already met problems involving the equilibrium of three forces which act through the same point. It can now be shown that all problems involving three forces in equilibrium fall into this category.

You can reason this in the following way. If the lines of action of two of the forces meet at point A, the resultant of those two forces is a force R passing through A. For equilibrium, the third force in the system must be $-R$, and its line of action must also pass through A. If it did not, the system would generate a turning effect and so would not be in equilibrium. The three forces therefore act through the same point.

Hence, you can state

> When a system of three forces is in equilibrium, the lines of action of the forces must all pass through a single point.

M2

Example 11

A uniform rod AB of weight W rests with end B against a smooth vertical wall. End A is hinged to horizontal ground. The rod is inclined at 45° to the horizontal.

Find the magnitude and direction of S, the reaction in the hinge, and the magnitude of the reaction R between the rod and the wall.

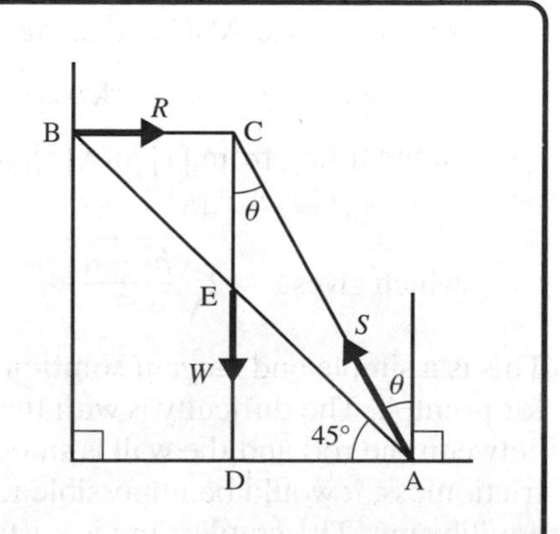

The reaction S at A makes an angle θ with the vertical, as shown. The lines of action of R and of W meet at C. As there are only three forces, the line of action of S must also pass through C. You can see that BCE and ADE are congruent right-angled isosceles triangles, which means that CD = 2AD. Therefore,

$$\tan\theta = \frac{AD}{CD} = 0.5 \quad \text{which gives} \quad \theta = 26.6°$$

You can now draw a triangle of forces, as shown on the right.
You know that $\tan\theta = 0.5$, and hence $R = \frac{1}{2}W$
From Pythagoras's theorem, you can find

$$S = \sqrt{W^2 + R^2} = \frac{W\sqrt{5}}{2}$$

A modelling problem

It was a criticism of old mechanics books and specifications that questions were set that made unrealistic assumptions in order to achieve an 'elegant solution', but which may not have been much use practically. You can work through this question from an old mechanics book, and then explore experimentally a more realistic model of the situation to look at the differences. The question is:

> A uniform rod AB of length $2a$ and weight W rests with A in contact with a smooth vertical wall. B is attached by means of a light inextensible string of length $2b$ to a point C on the wall, a distance x vertically above A. If the system is in equilibrium, find the length AC in terms of a and b.

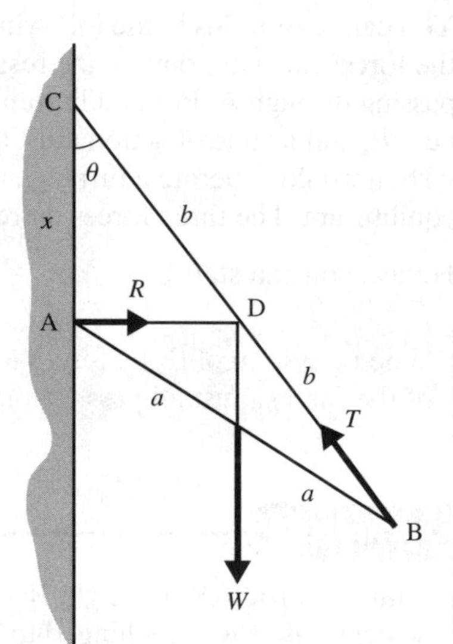

The solution to this problem uses the principle that three forces in equilibrium act through the same point.

The wall is smooth, so the reaction, R, is normal. W, R and the tension, T, all go through at point D, the mid-point of BC.
From triangle ACD, you have

$$x = b\cos\theta \qquad [1]$$

From triangle ABC and using the cosine rule you have

$$4a^2 = x^2 + 4b^2 - 4xb\cos\theta \qquad [2]$$

Substituting from [1] into [2], you find

$$4a^2 = x^2 + 4b^2 - 4x^2$$

which gives $x = 2\sqrt{\dfrac{b^2 - a^2}{3}}$

This is a simple and elegant solution which leads to a unique position for point A. The difficulty is with the assumption that the contact between the rod and the wall is smooth. If the wall were really frictionless, it would be impossible to get the rod to remain in equilibrium. The equilibrium is unstable, which means that once the system has made the slightest move from the equilibrium position (and any vibration or air current would achieve this), it would continue to move until both A and B were against the wall.

A more realistic scenario
You are going to find a model that is more closely related to the real world. You will still make assumptions, such as ignoring any weight or stretch in the string, but clearly you cannot ignore friction.

If the wall contact is rough, there is a range of values of x for which the system is in equilibrium. There is an upper limiting position in which A is about to slide **up** the wall, and a lower limiting position in which it is about to slide **down**.

The problem can be more realistically stated as:

A uniform rod AB of length $2a$ and weight W rests with A in contact with a rough vertical wall, with the coefficient of friction μ. B is attached by means of a light inextensible string of length $2b$ to a point C on the wall, a distance x vertically above A. The angle BCA is θ. Find expressions for μ for both the upper and lower limiting equilibrium positions.

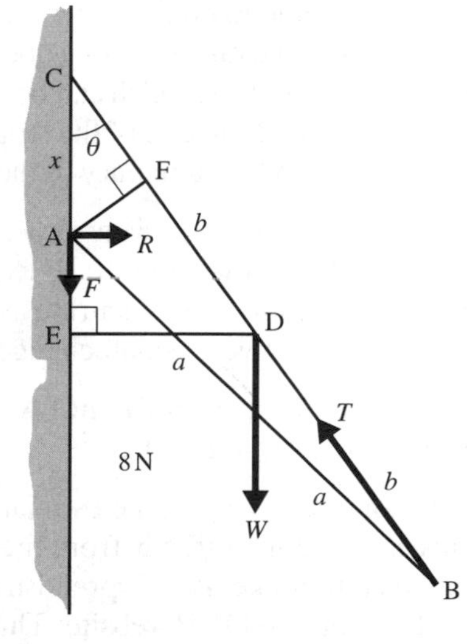

The theory
In the upper limiting position, as A is about to slide up the wall, the friction force F acts downwards.

In the experiment the value you will measure is x. From this you can then calculate θ from triangle ABC using the cosine rule:

$$\cos\theta = \frac{x^2 + 4b^2 - 4a^2}{4bx}$$

From triangle CDE, you have $DE = b\sin\theta$.
From triangle CAF, you have $AF = x\sin\theta$.

Resolving horizontally, you have $R = T\sin\theta$ [1]
Resolving vertically, you have $T\cos\theta = F + W$ [2]

Taking moments about A, you have $W \times DE = T \times AF$
which gives $Wb\sin\theta = Tx\sin\theta$
and hence $Wb = Tx$ [3]
Friction is limiting, and so $F = \mu R$ [4]

Remember: For a body that is about to move, $F = \mu R$ where μ is the coefficient of friction.

Substituting from [1] and [4] into [2], you find
$T\cos\theta = \mu T\sin\theta + W$ [5]

Substituting for W from [3] into [5], you find

$$T\cos\theta = \mu T\sin\theta + \frac{Tx}{b}$$

and hence $\mu = \dfrac{b\cos\theta - x}{b\sin\theta}$

You can draw the diagram for the lower limiting equilibrium position and confirm that in this case

$$\mu = \frac{x - b\cos\theta}{b\sin\theta}.$$

So, in each case by measuring x you will be able to calculate θ and hence an estimate of μ.

The experiment

There are two obvious ways in which you can test this theory by carrying out an experiment with a suitable rod and wall.

1 Calculate estimates of μ from each limiting equilibrium position and compare the results.
2 Calculate μ from the upper limiting equilibrium position. Use your result to predict the value of x corresponding to the lower limiting equilibrium position and confirm this experimentally.

When you come to do the experiment, you will have more success if you replace the rod with a strip of wood 2 or 3 cm wide, and the string with a fairly wide, light tape which can be stuck to the underside of the strip at B, passing round the end of the strip, and pinned to the wall at C. This helps to prevent the tendency of the rod to slip sideways on the wall.

You should find it relatively easy to gradually move the rod into a position where A is about to slide upwards. Measure the distance AC. You need to be as accurate as possible, because the formula for μ is quite sensitive to small changes in the input values.

M2

Now move the strip (rod) until A is about to slide downwards. Again measure the distance AC.

You can now calculate the estimated values of μ from your results. This can be done directly from the formulae above. Alternatively, you may wish to make use of spreadsheet MOMENTS2, which is available on the OUP website. This will be particularly helpful for the second part of the problem when you try to find x from a given value of μ.

Just type in the address: http://www.oup.co.uk/secondary/mechanics

When the author did this experiment, he used a 40 cm rod and a 54 cm tape, so that $a = 0.2$ m and $b = 0.27$ m. His results were:

Upper position: $x = 0.194$ m, giving $\theta = 36.13°$
Lower position: $x = 0.233$ m, giving $\theta = 42.38°$

1 Taking the first approach, that of calculating two estimates of μ and comparing them, gives

$$\text{Estimate from upper position } \mu = \frac{0.27\cos 36.13° - 0.194}{0.27\sin 36.13°}$$
$$= 0.15$$

$$\text{Estimate from lower position } \mu = \frac{0.233 - 0.27\cos 42.38°}{0.27\sin 42.38°}$$
$$= 0.18$$

These two estimates of μ are in reasonable agreement, although a detailed analysis of errors would be needed to decide whether they do indeed support the model.

2 Instead of finding a second estimate of μ using the lower position, the second approach uses the estimated μ (0.15 from the above data) from the upper position to predict x for the lower position.

If you are using the spreadsheet, this can be done by entering the value for μ and using the 'goal seek' command to find x. Otherwise, you will need to solve the equations

$$\cos\theta = \frac{x^2 + 4b^2 - 4a^2}{4bx} \qquad [1]$$

$$\text{and} \quad \mu = \frac{x - b\cos\theta}{b\sin\theta} \qquad [2]$$

Substituting known values ($a = 0.2$, $b = 0.27$, $\mu = 0.15$) gives:

From [1]: $x^2 - 1.08\,x\cos\theta + 0.1316 = 0$ [3]
From [2]: $x = 0.0405\sin\theta + 0.27\cos\theta$ [4]

Substituting from [4] into [3], gives

$$0.001\,640\,025\sin^2\theta - 0.021\,87\sin\theta\cos\theta - 0.2187\cos^2\theta + 0.1316 = 0$$

Dividing through by $\cos^2\theta$ and using $\sec^2\theta = 1 + \tan^2\theta$, gives

$$0.133\,240\,25\tan^2\theta - 0.021\,87\tan\theta - 0.0871 = 0$$

Solving this quadratic equation:

$$\tan\theta = 0.8947 \text{ (or } -0.7306\text{) and hence } \theta = 41.82°$$

Substituting in [4] leads to $x = 0.228$ m, which is in good experimental agreement with the recorded value of 0.233 m.

The error is about 2.5%, which is a good experimental result.

Exercise 2D

1 A uniform rod AB of length 3 m and mass 15 kg is hinged at A. A light string is attached to B and holds the rod in equilibrium at an angle of 60° to the upward vertical through A. Find the tension in the string when

a) the string is at right angles to AB
b) the string is vertical
c) the string is horizontal.

2 The diagram shows a horizontal uniform rod AB of length $2a$ and weight W. A light string is attached to A and B and passes through a smooth ring at C, vertically above A, so that angle ABC is 30°. A horizontal force P is applied to A.

a) Show that the system cannot be in equilibrium.
b) A weight W is attached at a point D on AB so that equilibrium is maintained. Find the distance AD and the force P.

3 The diagram shows a cross-section of a uniform horizontal shelf hinged to a vertical wall. The length AB is $2a$ and the shelf has weight W. The shelf is supported by a light string CD connecting a point D on the shelf, where AD $= x$, to a point C on the wall, a distance a vertically above A. The breaking strain of the string is $4W$. Find the minimum value of x.

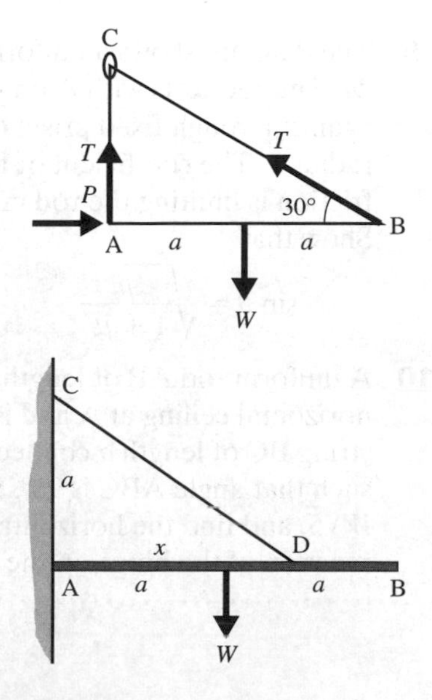

4 A uniform rod AB of mass 10 kg and length 2 m rests with A on smooth horizontal ground and B on a smooth peg 1 m above the ground. The rod is held in position by a horizontal force of P N at A. Find the value of P and the magnitude of the reactions at A and B.

5 A uniform ladder of mass 20 kg and length 3 m rests against a smooth wall with the bottom of the ladder on smooth horizontal ground and attached by means of a light inextensible string, 1 m long, to the base of the wall.

a) Find the tension in the string.

b) If the breaking strain of the string is 250 N, find how far up the ladder a man of mass 80 kg can safely ascend.

6 A uniform ladder of weight W rests at an angle α to the horizontal with its top against a smooth vertical wall and its base on rough horizontal ground with coefficient of friction 0.25. Find the minimum value of α if the ladder does not slip.

M2

7 A uniform ladder of weight W rests at an angle α to the horizontal with its top against a rough vertical wall and its base on rough horizontal ground, with coefficient of friction 0.25 at each contact. Find the minimum value of α if the ladder does not slip.

8 The diagram shows a cross-section ABCD of a uniform rectangular block of mass 20 kg. AB is 0.75 m and BC is 1 m. The block rests with A on rough horizontal ground and AB at 20° to the horizontal. It is held in place by a horizontal force P N applied at C. The block is on the point of slipping. Find the value of P and the coefficient of friction between the block and the ground.

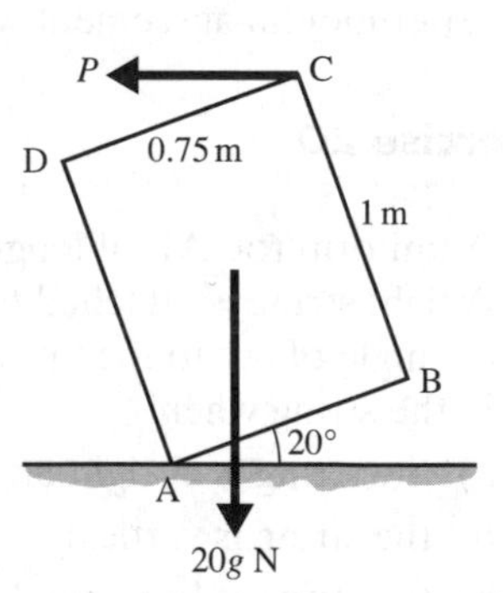

9 The diagram shows a uniform rod AB of weight W and length $2a$. The rod rests with A on rough horizontal ground and leans against a rough fixed prism of semicircular cross-section of radius a. The coefficient of friction at both contacts is μ. When friction is limiting the rod makes an angle θ with the horizontal. Show that

$$\sin\theta = \sqrt{\frac{\mu}{1+\mu^2}}$$

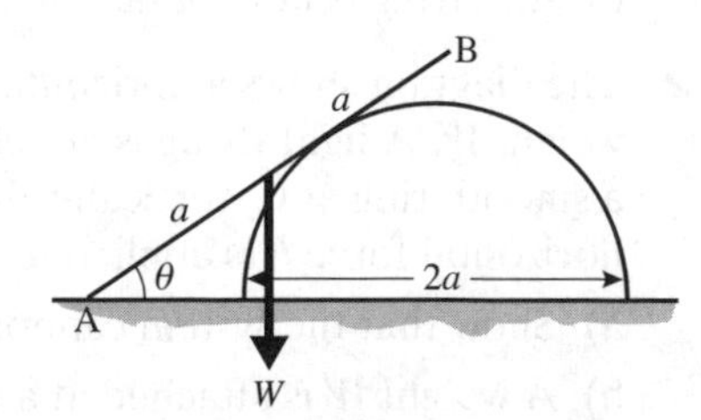

10 A uniform rod AB of length $2a$ and weight W is hinged to a horizontal ceiling at A and is suspended by a light inextensible string BC of length a connecting B to a point C on the ceiling such that angle ABC is 90°. Show that the tension in the string is $W\sqrt{5}$, and find the horizontal and vertical components of the reaction of the hinge on the rod.

Summary

You should know how to ...	Check out
1 Find the moment of a force about a given point.	**1** A force of 10 N acts as shown. Find the moment of the force about the point A.
2 Know the conditions for a rigid body to be in equilibrium.	**2** State two conditions necessary for a rigid body to be in equilibrium.
3 Calculate unknown forces acting on a rigid body in equilibrium.	**3** A uniform ladder, of mass 20 kg, rests against a smooth vertical wall. The ladder is inclined at 10° to the vertical and the foot of the ladder is on rough, horizontal ground. Find the reaction force between the ladder and the wall.

Revision exercise 2

1 A uniform plank, AB, rests horizontally on two fixed vertical supports at C and D. The plank has mass 10 kg and length 2.5 m. The supports at C and D are 0.25 m from A and B respectively, as shown in the diagram.

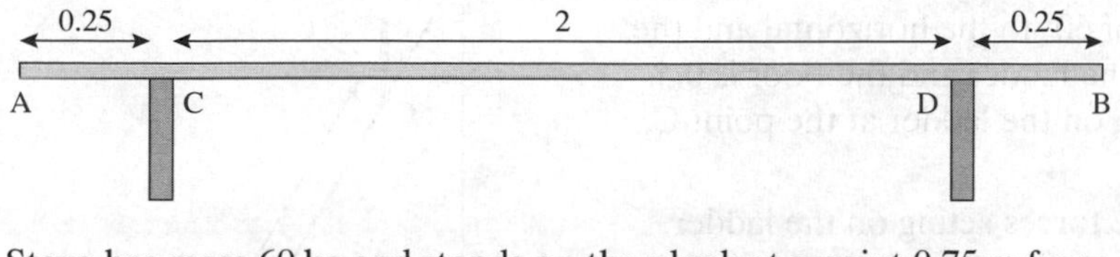

Steve has mass 60 kg and stands on the plank at a point 0.75 m from A.

a) Draw a diagram showing all the forces acting on the plank.

b) Find the reaction of the supports on the plank. *(AQA, 2002)*

2 A uniform rod, AB, has length 4 m, and C is the mid-point of AB.

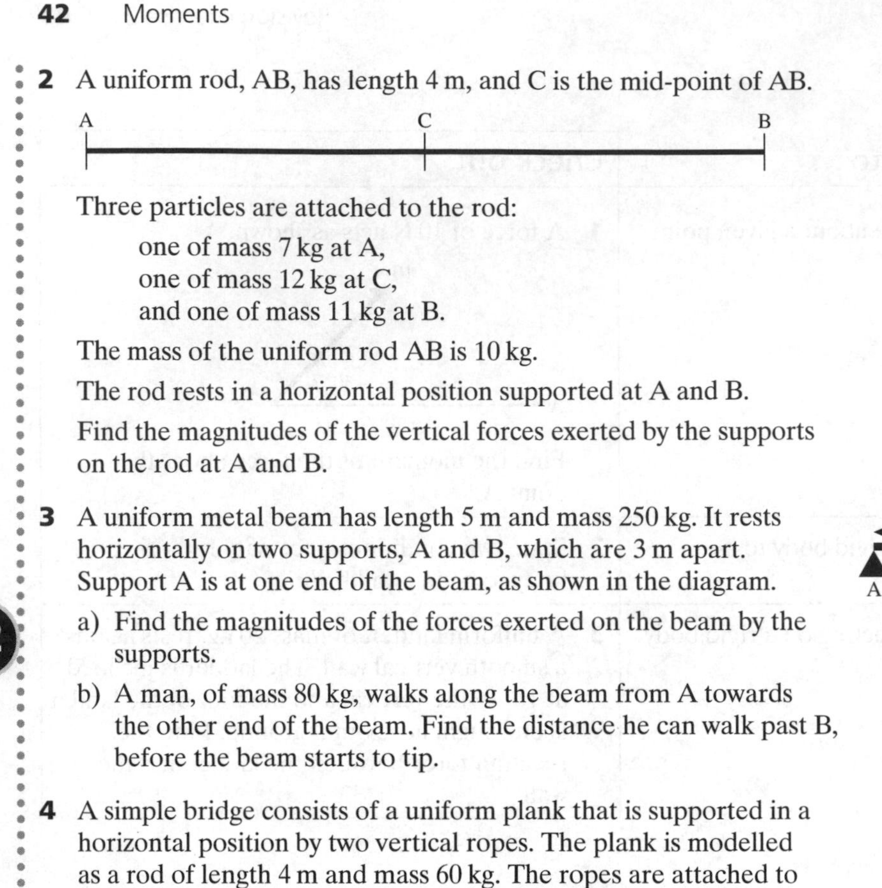

Three particles are attached to the rod:

one of mass 7 kg at A,
one of mass 12 kg at C,
and one of mass 11 kg at B.

The mass of the uniform rod AB is 10 kg.

The rod rests in a horizontal position supported at A and B.

Find the magnitudes of the vertical forces exerted by the supports on the rod at A and B. *(AQA, 2003)*

3 A uniform metal beam has length 5 m and mass 250 kg. It rests horizontally on two supports, A and B, which are 3 m apart. Support A is at one end of the beam, as shown in the diagram.

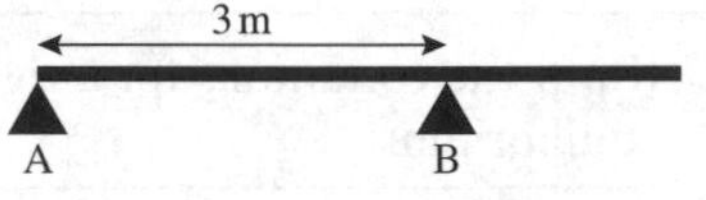

a) Find the magnitudes of the forces exerted on the beam by the supports.

b) A man, of mass 80 kg, walks along the beam from A towards the other end of the beam. Find the distance he can walk past B, before the beam starts to tip. *(AQA, 2002)*

4 A simple bridge consists of a uniform plank that is supported in a horizontal position by two vertical ropes. The plank is modelled as a rod of length 4 m and mass 60 kg. The ropes are attached to the plank, at A and B, as shown in the diagram.

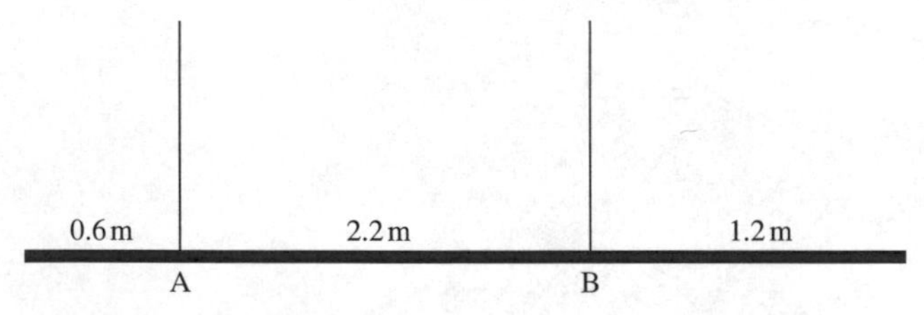

a) Draw a diagram to show the three forces acting on the plank.

b) Show that the tension in the rope, attached to the plank at A, is 214 N, correct to 3 significant figures.

c) Find the tension in the rope attached to the plank at B. *(AQA, 2002)*

5 A uniform ladder AB, of mass 20 kg and length 5 m, rests with B on a rough horizontal floor and with A against a smooth vertical wall. The ladder is at an angle of 60° to the horizontal and the coefficient of friction between the ladder and the floor is 0.4. A man of mass 80 kg is standing on the ladder at the point C, x metres from B.

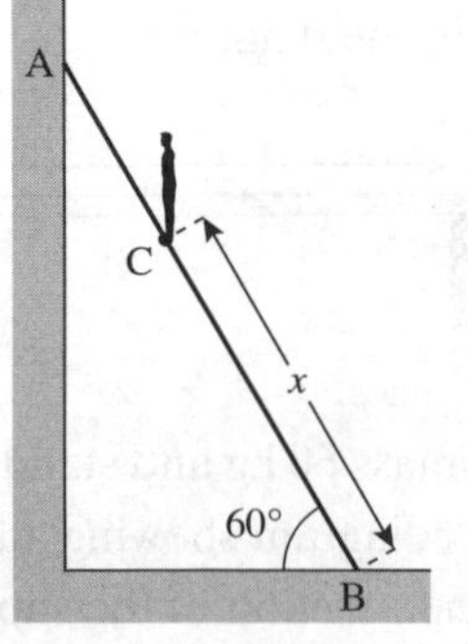

a) Draw a diagram showing the forces acting on the ladder.

b) Given that the ladder is in limiting equilibrium:

i) show that the force on the ladder at A is of magnitude 392 N;

ii) find the value of x. *(AQA, 2003)*

6 The diagram shows a uniform ladder AB of length 4 m and mass 10 kg. The ladder rests with one end A in contact with a smooth vertical wall, and the other end B in contact with a rough horizontal floor. The coefficient of friction between the ladder and the floor is 0.3. When a decorator, of mass 70 kg, stands at the point C on the ladder, where BC = 3 m, the ladder is on the point of slipping.

a) Draw a diagram showing the forces acting on the ladder when the decorator is standing at the point C.

b) Determine the angle the ladder makes with the horizontal, giving your answer in degrees to one decimal place.

(*AQA, 2002*)

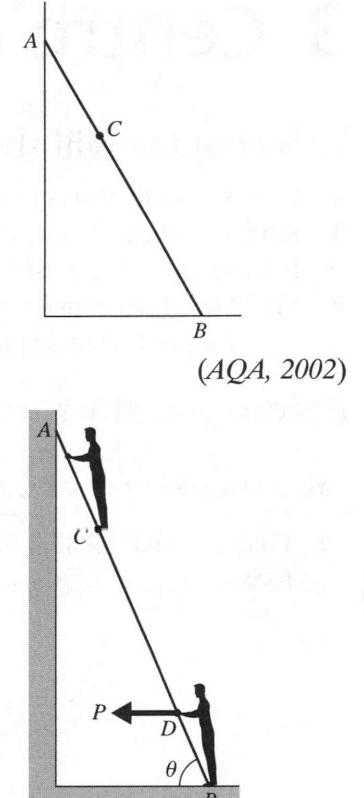

7 A uniform ladder AB, of weight W and length $4a$, is resting with A against a smooth vertical wall and B on rough horizontal ground. A man of weight $4W$ is standing at a point C on the ladder, where BC = $3a$. The force exerted by the man's hands on the ladder is negligible and the centre of mass of the man is vertically above C.

Another man stands on the ground near the bottom of the ladder and pushes on the ladder at the point D, where BD = a. The force applied at D is horizontal, of magnitude P and is just sufficient to prevent the ladder sliding away from the wall.

The ladder is inclined at an angle $\theta = \tan^{-1}\left(\frac{12}{5}\right)$ to the horizontal.

The coefficient of friction between the base of the ladder and the ground is $\frac{1}{4}$.

a) Show that the horizontal frictional force at B is of magnitude $\frac{5W}{4}$.

b) Find the value of P in terms of W.

(*AQA, 2004*)

M2

3 Centre of mass

This chapter will show you how to

- ✦ Understand the concept of a centre of mass
- ✦ Find the centre of mass of a system of particles
- ✦ Find the centre of mass of a simple uniform body by symmetry
- ✦ Find the centre of mass of a composite body
- ✦ Find the position of a body when it is suspended in equilibrium

Before you start

You should know how to ...	Check in
1 Find the total moment of a system of parallel forces.	**1** A light rod AB of length 4 m rests in a horizontal position on supports at A and B. Particles of mass 2 kg, 4 kg and 6 kg are attached to the rod at 1 m, 2 m and 3 m from A respectively. Find the total moment of their weights about A and hence find the reaction in the support at B.

An introductory experiment

Take a sheet of card and cut it to an irregular shape. Make a number of holes at random positions around its edge.

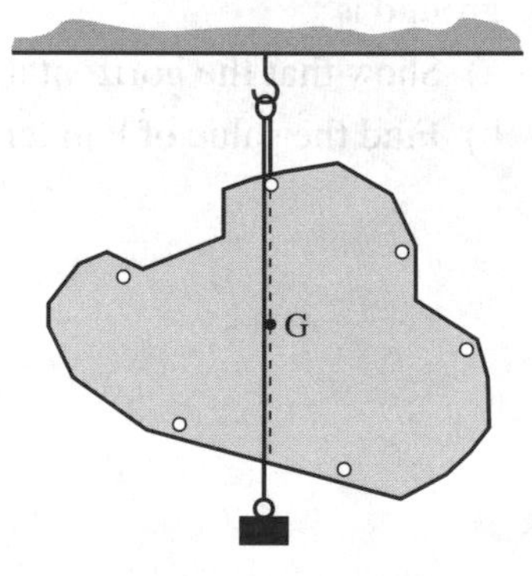

Take a length of string. Tie a loop about two thirds of the way along. Tie a small weight to the end of the longer portion, and the other end to one of the holes in the card.

Suspend the card and weight on the loop. When it is hanging at rest, mark on the card the line indicated by the string.

Repeat the process with the card attached by a different hole. The two lines you have drawn will cross at a point G.

Now suspend the card from the other holes in turn. You will find that the line of the string always passes through G.

Interpretation

The string with the weight gives a vertical line through the point of suspension. The line you drew was the **line of action** of the weight of the card, which must pass through the point of suspension as the card is in equilibrium.

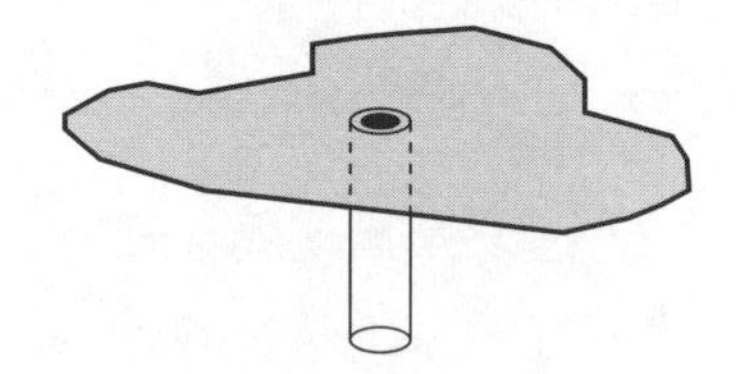

As this line of action always passes through G, the weight of the card is exactly the same as if the card were a single point mass positioned at G.

You should be able to balance the card on the flat end of a pencil placed at G.

3.1 Centre of gravity, centre of mass and centroid

You often meet these three terms used almost interchangeably. Strictly speaking, the point you found in the experiment is the **centre of gravity**, which can be defined as follows:

> When a system consisting of one or more bodies having a total mass M is acted upon by gravity, there is a point G such that the magnitude and line of action of the weight of the system are the same as those of a particle of mass M placed at G.
> G is the **centre of gravity** of the system.

If the object under consideration is small, the centre of gravity is independent of the orientation of the object and coincides with a point called the **centre of mass**. This depends on the distribution of mass within the object. Even in the weightless conditions of space, where centre of gravity is meaningless, the centre of mass is still important as it affects, for example, the behaviour of the object in collision with other objects.

The orientation means the position of the object relative to a fixed point, for example, the angle it makes with a plane.

M2

The two points coincide when the object is small enough for the acceleration due to gravity to have effectively the same magnitude and direction at each point of the object. This will be the case in all situations you will encounter in this course.

See the appendix to this book (page 170) for an exploration of a situation where this is not the case.

Consider a simple system consisting of two point masses of weight w_1 and w_2. They are placed at points A and B on a horizontal line through an origin O so that $OA = x_1$ and $OB = x_2$.

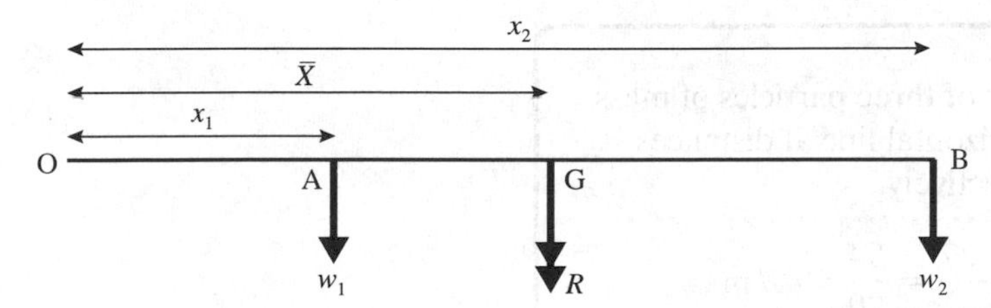

Assume that the system is small enough to consider the weights as parallel forces. They are then equivalent to a resultant force, R, acting through some point G on AB, where $OG = \overline{X}$.

The principle of moments, discussed in Chapter 2, shows that the moment of R about O is the same as the total moments of w_1 and w_2 about O. That is,

$$R\overline{X} = w_1x_1 + w_2x_2$$

and hence $\overline{X} = \dfrac{w_1x_1 + w_2x_2}{R}$

But $R = w_1 + w_2$, which gives

$$\overline{X} = \frac{w_1x_1 + w_2x_2}{w_1 + w_2}$$

More generally, the line OAB could be at an angle θ to the horizontal. The above result still holds, but the working changes as follows.

$$R\bar{X}\cos\theta = w_1x_1\cos\theta + w_2x_2\cos\theta$$

and hence $\bar{X} = \dfrac{w_1x_1 + w_2x_2}{R}$

But $R = w_1 + w_2$, which gives

$$\bar{X} = \frac{w_1x_1 + w_2x_2}{w_1 + w_2}$$

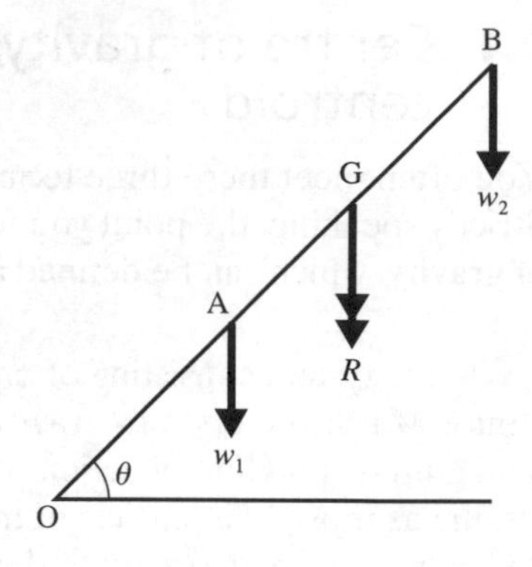

The position of G is therefore independent of the orientation of the system.

G is both the **centre of gravity** and the **centre of mass** of the system. You can define the centre of mass of an object as follows:

> The **centre of mass** of a system is the centre of gravity of that system when it is placed in a gravitational field such that each part of the system is subject to the same gravitational acceleration.

M2

For a solid or a lamina (plane shape) there is a third centre, the **centroid**, which is a geometrical centre. For example, the centroid of a rectangular lamina is at the intersection of its diagonals, P. The centroid coincides with the centre of mass when the object is made of a uniformly dense material. The centroid is important in certain cases, for example, when a plane surface is subjected to the pressure of a uniform air flow. Regardless of the mass distribution of the object, the resultant force caused by the pressure acts through the centroid of the surface.

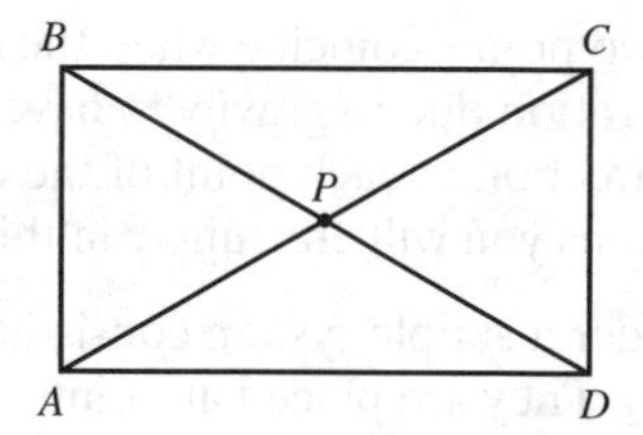

Example 1

Find the position of the centre of mass of three particles of mass 4 kg, 10 kg and 6 kg, which lie on a horizontal line at distances 3 m, 4 m and 7 m from a point A, respectively.

$$\bar{X} = \frac{4\times3 + 10\times4 + 6\times7}{4 + 10 + 6} = \frac{94}{20} = 4.7\text{ m}$$

The centre of mass is on the horizontal line, 4.7 m from A.

Finding the centre of mass of a system of point masses

You need to be able to find the centre of mass of a system of particles independent of a particular gravitational field.

Suppose you have a system consisting of particles $m_1, m_2, \dots, m_n$ placed at points $(x_1, y_1), (x_2, y_2), \dots, (x_n, y_n)$ in a plane, as shown.

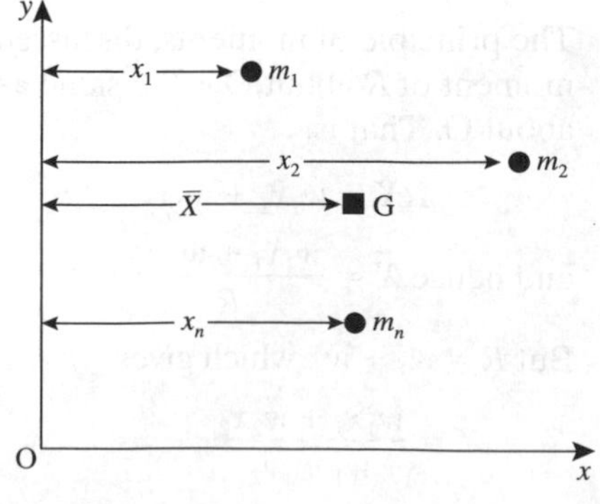

If the system is placed in a uniform gravitational field perpendicular to the plane, the resultant weight is the total of the weights of the individual particles and acts through the centre of mass, $G(\bar{X}, \bar{Y})$.

Taking moments about the y-axis, you find

$$(m_1g + m_2g + \dots m_ng)\overline{X} = m_1gx_1 + m_2gx_2 + \dots + m_ngx_n$$

and hence

$$\overline{X} = \frac{m_1gx_1 + m_2gx_2 + \dots + m_ngx_n}{m_1g + m_2g + \dots + m_ng}$$

Cancelling by g, you get

$$\overline{X} = \frac{m_1x_1 + m_2x_2 + \dots + m_nx_n}{m_1 + m_2 + \dots + m_n}$$

This can be expressed more easily using Σ notation:

$$\overline{X} = \frac{\sum_{i=1}^{n} m_ix_i}{\sum_{i=1}^{n} m_i} \quad \text{or} \quad \overline{X}\sum_{i=1}^{n} m_i = \sum_{i=1}^{n} m_ix_i$$

You can repeat this analysis, taking moments about the x-axis, to get

$$\overline{Y} = \frac{\sum_{i=1}^{n} m_iy_i}{\sum_{i=1}^{n} m_i} \quad \text{or} \quad \overline{Y}\sum_{i=1}^{n} m_i = \sum_{i=1}^{n} m_iy_i$$

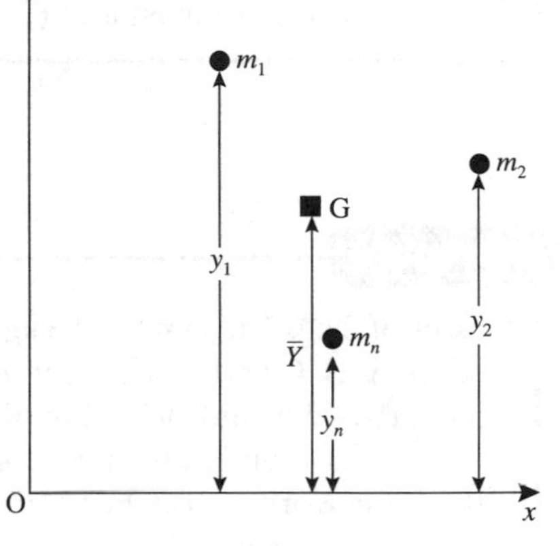

It is possible to combine the above formulae into a vector equation:

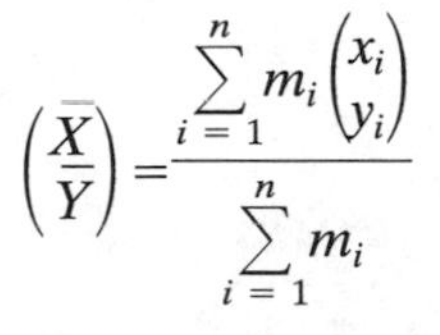

$$\begin{pmatrix}\overline{X}\\ \overline{Y}\end{pmatrix} = \frac{\sum_{i=1}^{n} m_i\begin{pmatrix}x_i\\ y_i\end{pmatrix}}{\sum_{i=1}^{n} m_i}$$

This can be extended to three dimensions as

$$\begin{pmatrix}\overline{X}\\ \overline{Y}\\ \overline{Z}\end{pmatrix} = \frac{\sum_{i=1}^{n} m_i\begin{pmatrix}x_i\\ y_i\\ z_i\end{pmatrix}}{\sum_{i=1}^{n} m_i}$$

This vector form can be expressed more neatly. If each mass m_i has position vector $\mathbf{r}_i$ and the centre of mass, G, has position vector $\overline{\mathbf{R}}$, then

$$\overline{\mathbf{R}} = \frac{\sum_{i=1}^{n} m_i\mathbf{r}_i}{\sum_{i=1}^{n} m_i}$$

Example 2

Masses of 2 kg, 3 kg and 5 kg are placed at A(3, 1), B(5, 7) and C(1, −4) respectively. Find the position of the centre of mass.

Taking moments about the y-axis

$$\bar{X} = \frac{2 \times 3 + 3 \times 5 + 5 \times 1}{2 + 3 + 5} = 2.6$$

Taking moments about the x-axis

$$\bar{Y} = \frac{2 \times 1 + 3 \times 7 + 5 \times -4}{2 + 3 + 5} = 0.3$$

So the centre of mass is at (2.6, 0.3).

You could combine these calculations using the vector form.

$$\bar{\mathbf{R}} = \frac{2\begin{pmatrix}3\\1\end{pmatrix} + 3\begin{pmatrix}5\\7\end{pmatrix} + 5\begin{pmatrix}1\\-4\end{pmatrix}}{2 + 3 + 5} = \begin{pmatrix}2.6\\0.3\end{pmatrix}$$

Again, the centre of mass is at (2.6, 0.3).

M2

Example 3

Masses of 2 kg, 4 kg, 5 kg and 3 kg are placed respectively at the vertices A, B, C and D of a light rectangular framework ABCD, where AB = 3 m and BC = 2 m. Further masses of 1 kg and 5 kg are placed at E and F, the mid-points of BC and CD respectively. If the framework is suspended from A, find the angle which AB makes with the vertical.

The mass of the rods forming the framework is assumed to be negligible in comparison with the masses attached to it.

Take AB and AD to be the x- and y-axes, as shown.

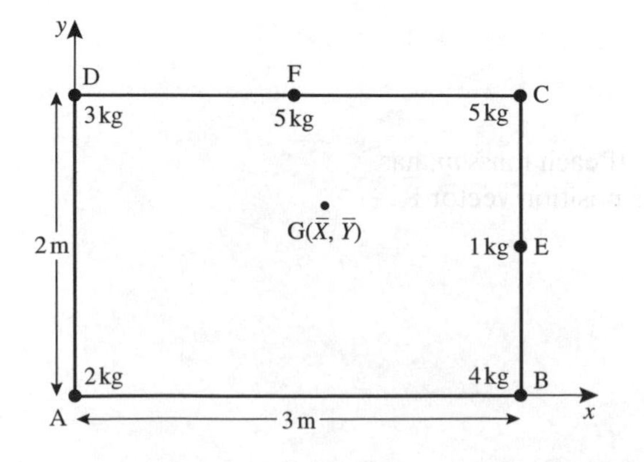

Let $G(\bar{X}, \bar{Y})$ be the centre of mass of the system.
It follows that

$$\bar{X} = \frac{2 \times 0 + 4 \times 3 + 1 \times 3 + 5 \times 3 + 5 \times 1.5 + 3 \times 0}{2 + 4 + 1 + 5 + 5 + 3} = 1.875 \text{ m}$$

$$\bar{Y} = \frac{2 \times 0 + 4 \times 0 + 1 \times 1 + 5 \times 2 + 5 \times 2 + 3 \times 2}{2 + 4 + 1 + 5 + 5 + 3} = 1.35 \text{ m}$$

Alternatively, in vector form

$$\begin{pmatrix} \bar{X} \\ \bar{Y} \end{pmatrix} = \frac{2\begin{pmatrix} 0 \\ 0 \end{pmatrix} + 4\begin{pmatrix} 3 \\ 0 \end{pmatrix} + 1\begin{pmatrix} 3 \\ 1 \end{pmatrix} + 5\begin{pmatrix} 3 \\ 2 \end{pmatrix} + 5\begin{pmatrix} 1.5 \\ 2 \end{pmatrix} + 3\begin{pmatrix} 0 \\ 2 \end{pmatrix}}{2 + 4 + 1 + 5 + 5 + 3}$$

So, the centre of mass is at G(1.875, 1.35) m.

When the framework is suspended from vertex A, the line AG is vertical, as shown.

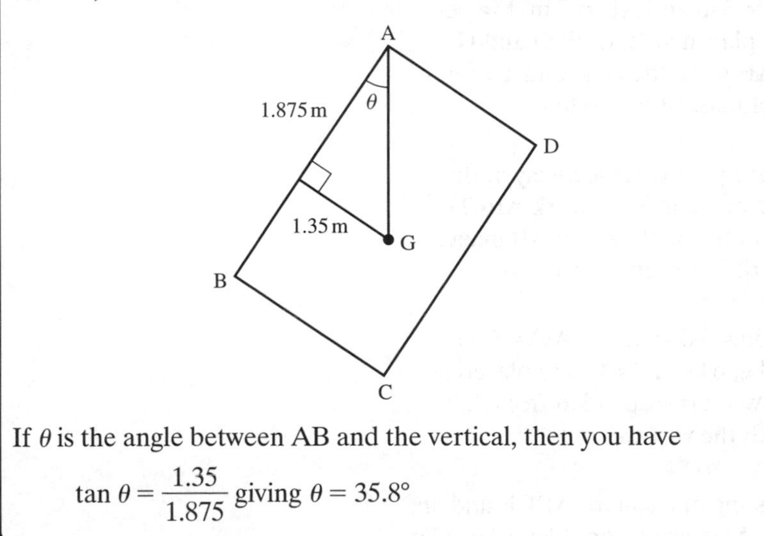

If θ is the angle between AB and the vertical, then you have

$$\tan \theta = \frac{1.35}{1.875} \text{ giving } \theta = 35.8°$$

Exercise 3A

1 Find the coordinates of the centre of mass of each of the following systems of masses placed respectively at the given points.

a) 3 kg, 5 kg and 7 kg at A(2, 5), B(3, 1) and C(4, 9)

b) 9 kg, 4 kg, 2 kg and 5 kg at A(4, 8), B(−2, 6), C(4, −4) and D(−2, −5)

c) 6 kg, 12 kg and 15 kg at A(0, −8), B(6, −3) and C(−4, −9)

d) 2 kg, 1 kg, 5 kg and 3 kg at A(2, 1, 6), B(3, 2, 0), C(5, −2, −8) and D(−1, 2, 3)

2 Masses of 3 kg, 8 kg and 5 kg are placed at points A, B and C with position vectors $3\mathbf{i} + 6\mathbf{j}$, $4\mathbf{i} - 2\mathbf{j}$ and $6\mathbf{i} - 8\mathbf{j}$ respectively. Find the position vector of the centre of mass.

3 Masses of 5 kg, 7 kg and 6 kg are placed at points A, B and C with position vectors $2\mathbf{i} - 7\mathbf{j} + 4\mathbf{k}$, $-3\mathbf{i} - 5\mathbf{j} + 8\mathbf{k}$ and $\mathbf{i} - 12\mathbf{k}$ respectively. Find the position vector of the centre of mass.

4 Masses of 4 kg, 9 kg and 6 kg are placed at A(5, 3), B(6, −2) and C(−1, 4) respectively. Where should a mass of 5 kg be placed so that the centre of mass of the whole system is at G(0, −1)?

5 A light rectangular framework ABCD has AB = 4 m and BC = 3 m. Masses of 5 kg, 4 kg, 2 kg and 3 kg are placed at A, B, C and D respectively. A fifth mass m kg is placed at a point E on CD so that the centre of mass of the system is at the centre of the rectangle. Find the value of m and the position of E.

6 A cuboidal framework of light rods has a rectangular base ABCD and vertices E, F, G and H vertically above A, B, C and D respectively, where AB = 4 m, AD = 3 m and AE = 2 m. Masses of 4 kg, 6 kg, 2 kg, 2 kg and 5 kg are placed at B, C, F, G and H respectively. Taking AB, AD and AE to be the x-, y- and, z-axes, find the coordinates of the centre of mass of the system.

7 Masses of 2 kg, 4 kg, 6 kg and 9 kg are placed respectively at the vertices A, B, C and D of a light rectangular framework ABCD, where AB = 5 m and BC = 3 m. Find the angle which AB makes with the vertical when the framework is suspended from A.

8 A light triangular framework ABC has AB = 4.6 m, AC = 6.3 m and angle BAC = 68°. Masses of 3 kg, 6 kg and 8 kg are placed at A, B and C respectively. The framework is suspended from A. Find the angle which AB makes with the vertical.

9 ABCDE is a light framework consisting of a square ABCE and an equilateral triangle CDE, as shown. Masses of 2 kg, 1 kg, 4 kg, 5 kg and m kg are attached to A, B, C, D and E respectively. The framework is then suspended from A. Find the value of m if the diagonal AC makes an angle of 20° with the vertical.

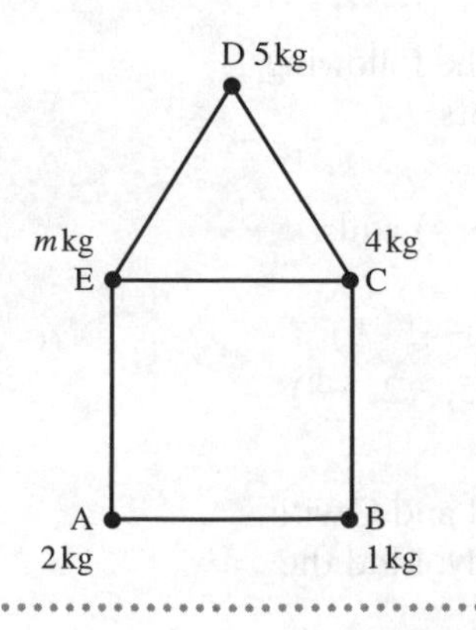

3.2 Centre of mass of a rigid body

You can find the centre of mass of some common shapes by considering their symmetry, provided that the bodies are uniformly dense.

This specification does not cover objects with variable density.

One dimension

A **uniform rod** can be modelled as a one-dimensional figure, whose thickness is assumed to be negligible compared with its length.

By symmetry, the centre of mass, G, of a rod or strip AB lies at the mid-point of AB.

Two dimensions

Any plane object whose thickness is negligible compared with its other dimensions is called a lamina.

Uniform rectangular lamina
By symmetry, the centre of mass, G, of a uniform rectangular lamina ABCD is at the intersection of its diagonals, as shown.

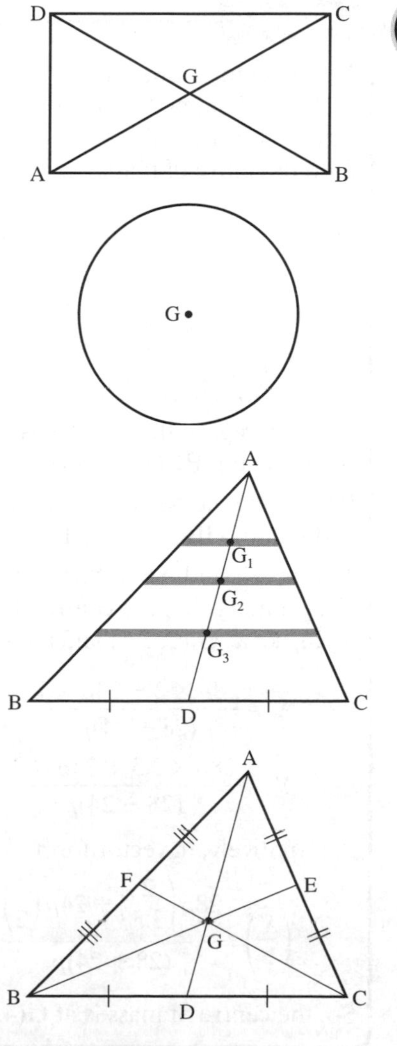

Uniform circular lamina
By symmetry, the centre of mass, G, of a uniform circular lamina is at the centre of the circle.

Uniform triangular lamina
A triangle can be regarded as being made up of a large number of strips of negligible thickness parallel to one of its sides, as shown. The centres of mass (G_1, G_2, G_3, etc) of these strips lie at their mid-points.

The centre of mass, G, of the triangle must therefore lie on the line formed by G_1, G_2, G_3, etc. This is the line AD in the diagram, joining A to the mid-point D of BC. This line is called a **median** of the triangle.

By considering the triangle divided into strips parallel to AC, you can see that G lies also on the median BE.

By considering the triangle divided into strips parallel to AB, you can see that G lies also on the median CF.

The medians of a triangle meet at the point which divides each median in the ratio 2 : 1. So, in this diagram, you have

$$AG : GD = BG : GE = CG : GF = 2 : 1$$

(You may not have encountered this standard geometrical result. A proof is given in the appendix to this book – page 171.)

Uniform semicircular lamina

The centre of mass of a uniform semicircular lamina of radius r lies on the line of symmetry, as shown, at a distance h from the straight edge (diameter).

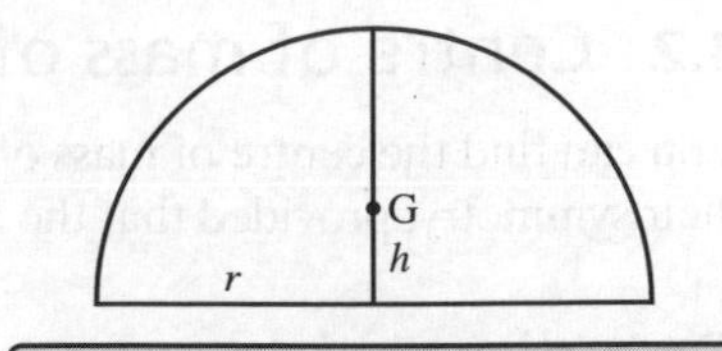

To find the centre of mass of this lamina it is necessary to use calculus methods. It can be shown that

$$h = \frac{4r}{3\pi}$$

Note The above section deals with one- and two-dimensional objects. You will not be required to know about the centre of mass of three-dimensional solids for the M2 examination.

Composite bodies

Many complex shapes are made up of several components, each of which is one of the standard shapes just described. Each of these components may be regarded as a point mass located at its centre of mass. The centre of mass of the overall shape may then be found from these point masses.

M2 **Example 4**

The diagram shows an L-shaped lamina ABCDEF with uniform density. Find its centre of mass.

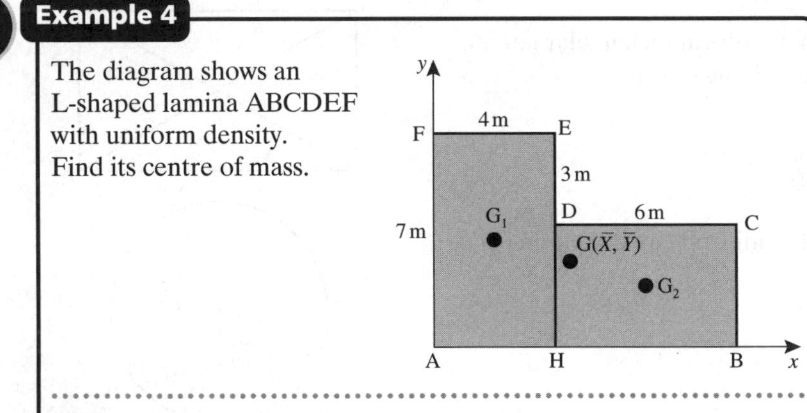

You can regard this lamina as being composed of two rectangles, AHEF and HBCD, as shown.

If you take the density to be ρ kg m^{-2}, the mass of AHEF is 28ρ kg and that of HBCD is 24ρ kg.

Taking AB and AF to be the x- and y-axes respectively, you can see by symmetry that the centre of mass of the body is that of a mass of 28ρ kg at $G_1(2, 3.5)$ and another of 24ρ kg at $G_2(7, 2)$.

$$\bar{X} = \frac{28\rho \times 2 + 24\rho \times 7}{(28 + 24)\rho} = 4.31$$

$$\bar{Y} = \frac{28\rho \times 3.5 + 24\rho \times 2}{(28 + 24)\rho} = 2.81$$

Alternatively, in vector form

$$\begin{pmatrix} \bar{X} \\ \bar{Y} \end{pmatrix} = \frac{28\rho \begin{pmatrix} 2 \\ 3.5 \end{pmatrix} + 24\rho \begin{pmatrix} 7 \\ 2 \end{pmatrix}}{(28 + 24)\rho} = \begin{pmatrix} 4.31 \\ 2.81 \end{pmatrix}$$

So, the centre of mass is at G(4.31, 2.81) m.

Notice that in Example 4 the density ρ cancelled out. This will always be the case for a uniform lamina, and hence for such cases you can assume that the density is 1. This assumption will be made from now on, and is equivalent to equating the mass with the area.

Example 5

The diagram shows a uniform lamina comprising a rectangle ABCD, with AB = 0.6 m and BC = 0.3 m, together with a semicircle with CD as diameter and a right-angled triangle BEC, where BE = 0.3 m. Taking AB and AD as the x- and y-axes, find the coordinates of the centre of mass of the lamina. Take the density to be 1 kg m^{-2}.

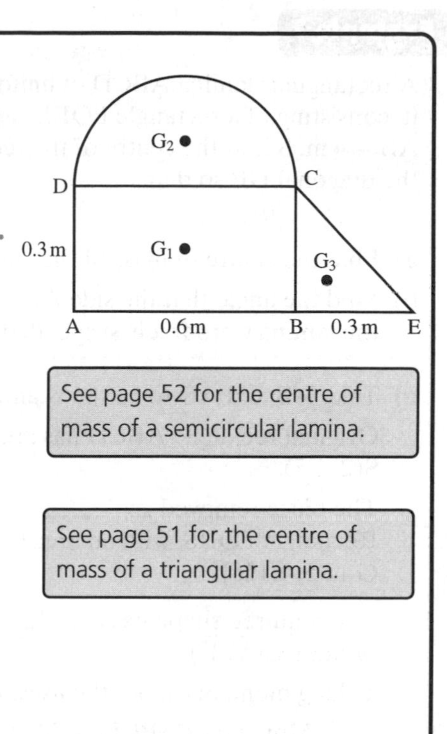

See page 52 for the centre of mass of a semicircular lamina.

See page 51 for the centre of mass of a triangular lamina.

The rectangle has area $0.3 \times 0.6 = 0.18$ m^2, and centre of mass $G_1(0.3, 0.15)$.

The semicircle has area $\frac{1}{2}\pi \times 0.3^2 = 0.141$ m^2, and centre of mass

$$G_2\left(0.3, 0.3 + \frac{4 \times 0.3}{3\pi}\right) = G_2(0.3, 0.427)$$

The triangle has area $\frac{1}{2} \times 0.3 \times 0.3 = 0.045$ m^2, and centre of mass $G_3(0.6 + \frac{1}{3} \times 0.3, \frac{1}{3} \times 0.3) = (0.7, 0.1)$

The whole lamina has area $0.18 + 0.141 + 0.045 = 0.366$ m^2.

Its centre of mass, $G(\overline{X}, \overline{Y})$,is given by

$$\overline{X} = \frac{0.18 \times 0.3 + 0.141 \times 0.3 + 0.045 \times 0.7}{0.366} = 0.349$$

$$\overline{Y} = \frac{0.18 \times 0.15 + 0.141 \times 0.427 + 0.045 \times 0.1}{0.366} = 0.251$$

Alternatively, in vector form

$$\begin{pmatrix}\overline{X}\\ \overline{Y}\end{pmatrix} = \frac{0.18\begin{pmatrix}0.3\\0.15\end{pmatrix} + 0.141\begin{pmatrix}0.3\\0.427\end{pmatrix} + 0.045\begin{pmatrix}0.7\\0.1\end{pmatrix}}{0.366} = \begin{pmatrix}0.349\\0.251\end{pmatrix}$$

So the centre of mass is G (0.349, 0.251) m.

M2

Example 6

The lamina in Example 4 is folded along DH so that the angle AHB is 90°, as shown. Find the centre of mass.

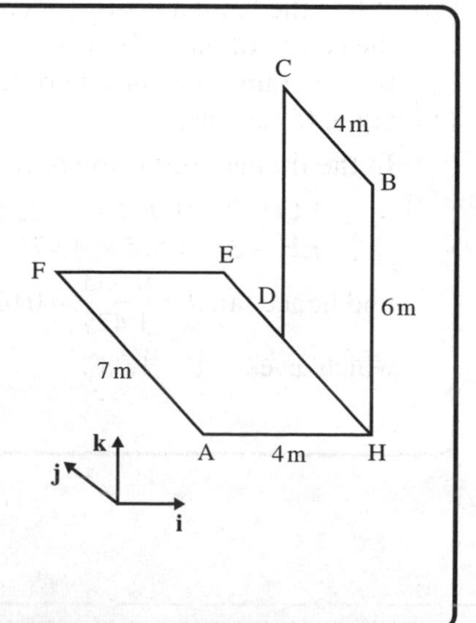

Taking A as the origin and directions **i**, **j** and **k** as shown, you have:

Centre of mass of AHEF is $G_1(2, 3.5, 0)$

Centre of mass of HBCD is $G_2(4, 2, 3)$

Therefore, the position, $\overline{\mathbf{R}}$, of the centre of mass of the whole body is given by

$$\overline{\mathbf{R}} = \frac{28\begin{pmatrix}2\\3.5\\0\end{pmatrix} + 24\begin{pmatrix}4\\2\\3\end{pmatrix}}{28 + 24} = \begin{pmatrix}2.92\\2.81\\1.38\end{pmatrix}$$

So, the centre of mass is at G (2.92, 2.81, 1.38) m.

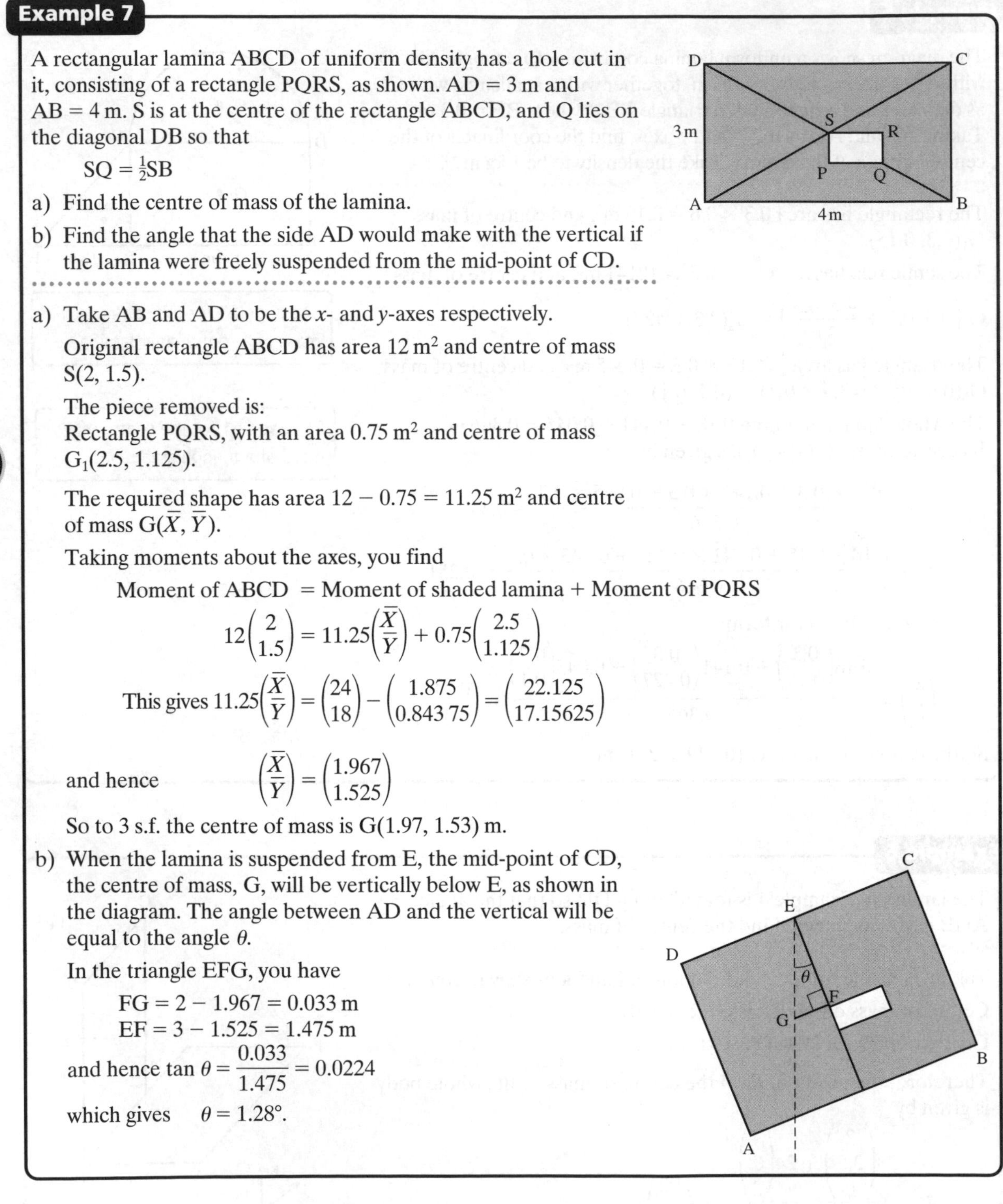

Example 7

A rectangular lamina ABCD of uniform density has a hole cut in it, consisting of a rectangle PQRS, as shown. AD = 3 m and AB = 4 m. S is at the centre of the rectangle ABCD, and Q lies on the diagonal DB so that

$$SQ = \tfrac{1}{2}SB$$

a) Find the centre of mass of the lamina.

b) Find the angle that the side AD would make with the vertical if the lamina were freely suspended from the mid-point of CD.

a) Take AB and AD to be the x- and y-axes respectively.

Original rectangle ABCD has area 12 m^2 and centre of mass S(2, 1.5).

The piece removed is:
Rectangle PQRS, with an area 0.75 m^2 and centre of mass $G_1(2.5, 1.125)$.

The required shape has area 12 − 0.75 = 11.25 m^2 and centre of mass $G(\overline{X}, \overline{Y})$.

Taking moments about the axes, you find

Moment of ABCD = Moment of shaded lamina + Moment of PQRS

$$12\begin{pmatrix}2\\1.5\end{pmatrix} = 11.25\begin{pmatrix}\overline{X}\\\overline{Y}\end{pmatrix} + 0.75\begin{pmatrix}2.5\\1.125\end{pmatrix}$$

This gives $11.25\begin{pmatrix}\overline{X}\\\overline{Y}\end{pmatrix} = \begin{pmatrix}24\\18\end{pmatrix} - \begin{pmatrix}1.875\\0.843\,75\end{pmatrix} = \begin{pmatrix}22.125\\17.15625\end{pmatrix}$

and hence $\begin{pmatrix}\overline{X}\\\overline{Y}\end{pmatrix} = \begin{pmatrix}1.967\\1.525\end{pmatrix}$

So to 3 s.f. the centre of mass is G(1.97, 1.53) m.

b) When the lamina is suspended from E, the mid-point of CD, the centre of mass, G, will be vertically below E, as shown in the diagram. The angle between AD and the vertical will be equal to the angle θ.

In the triangle EFG, you have

FG = 2 − 1.967 = 0.033 m
EF = 3 − 1.525 = 1.475 m

and hence $\tan\theta = \dfrac{0.033}{1.475} = 0.0224$

which gives $\theta = 1.28°$.

Exercise 3B

1 Find the centre of mass of each of the following laminae relative to the origin O and the axes shown. You may assume a uniform density of 1 kg m^{-2} in each case.

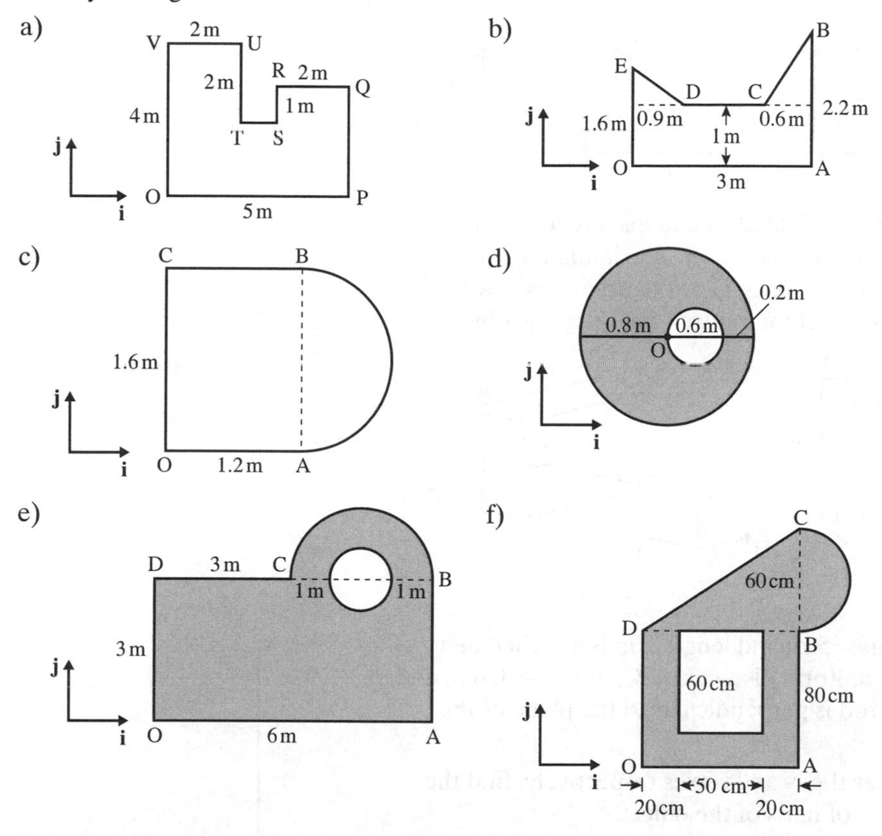

2 The diagram shows a triangular lamina ABC, in which angle ABC is 90°, AB = 0.6 m and BC = 0.9 m. The triangle is attached to a second (rectangular) lamina PQRS, where PQ = 1.2 m and PS = 0.8 m, so that BC lies on PQ and PB = 0.2 m, as shown.

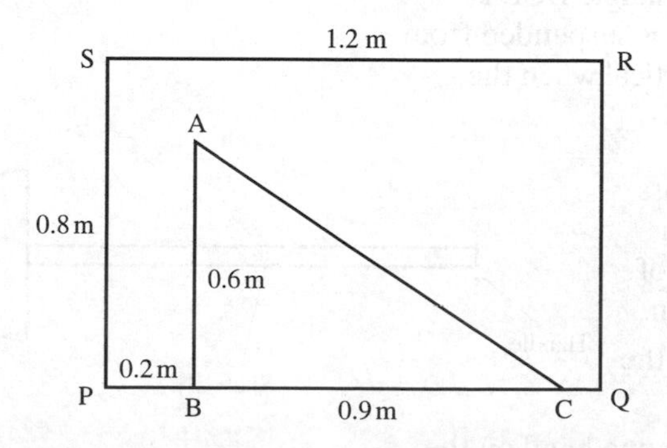

Assume that both laminae have uniform densities of 1 kg m^{-2}. Taking PQ and PS to be the x- and y-axes respectively, find the position of the centre of mass of the combined object.

3 A uniform rectangular card ABCD of density ρ kg m^{-2} is folded along OF and BE, as shown in the diagram. AB = 120 cm, AD = 40 cm, AO = 20 cm and CE = 40 cm. Taking O as the origin and axes as shown, find the centre of mass of the folded card.

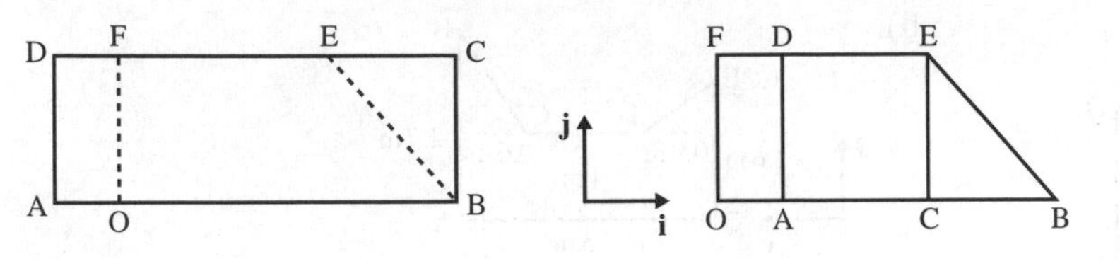

4 The lamina in the diagram is folded so that each rectangle is perpendicular to its neighbour, as shown. The lamina has a uniform density of ρ kg m^{-2}. Taking O as the origin and axes as shown, find the coordinates of the centre of mass of the object.

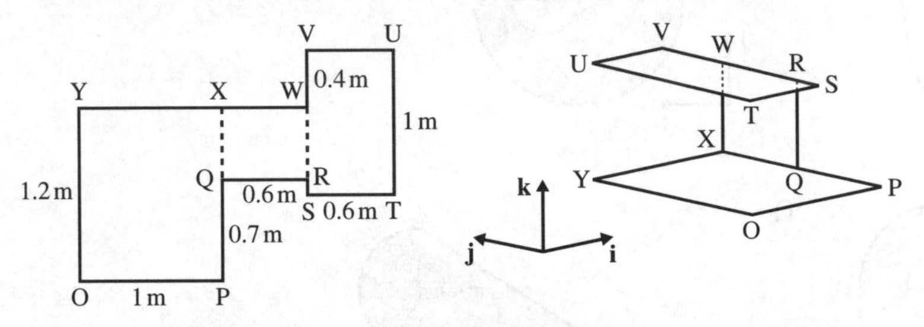

M2

5 A uniform rod AB of mass 5 kg and length 2 m is attached at a point B on the rim of a uniform disc, centre C, of radius 0.6 m and mass 10 kg, so that the rod is perpendicular to the plane of the disc, as shown.

a) Taking BC and BA as the x- and y-axes respectively, find the position of the centre of mass of the object.

b) If the object is suspended from A, find the angle between the rod AB and the vertical.

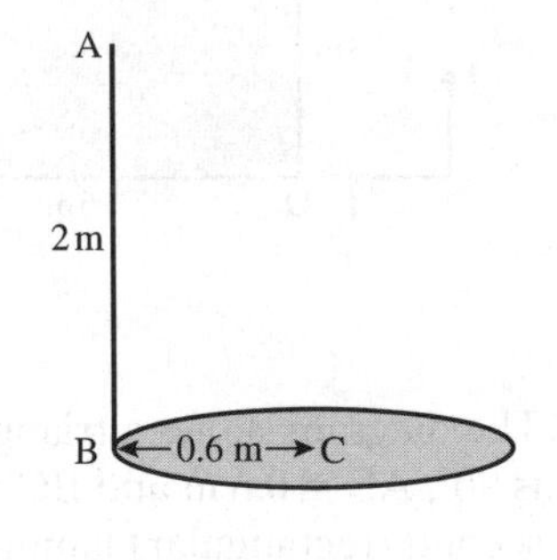

6 A uniform rectangular lamina ABCD has AB = 60 cm and BC = 30 cm. E is the mid-point of CD. The triangle BCE is removed from the lamina, and the remainder is suspended from E. Find the angle that AD makes with the vertical when the lamina hangs in equilibrium.

7 A lawn edging tool comprises a handle, of mass 30 g and negligible thickness, a shaft of length 1 m and mass 150 g, and a semicircular blade of diameter 20 cm and density 3 g cm^{-2}, as shown.

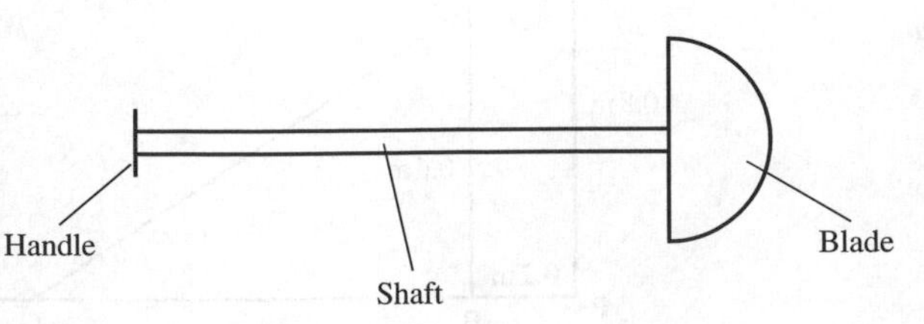

a) Find the distance of the centre of mass of the tool from the handle.

b) A gardener holds the tool horizontal, with one hand on the handle and the other at the mid-point of the shaft. Find the magnitude and direction of the force she must exert with each hand.

8 The diagram shows a chair made of four rods, AP, IQ, JR and DS, and two rectangular laminae ABCD and HIJK.
AP = DS = AD = 50 cm. IQ = JR = 120 cm. AB = HI = 40 cm.
The density of the rods is $600\ \text{g m}^{-1}$, and the density of each lamina is $4\ \text{kg m}^{-2}$.

a) Taking P as the origin, with PQ, PS and PA as the x-, y- and z-axes respectively, find the coordinates of the centre of mass of the chair.

b) Find the angle that the seat makes with the horizontal if the chair were suspended from the mid-point of IJ.

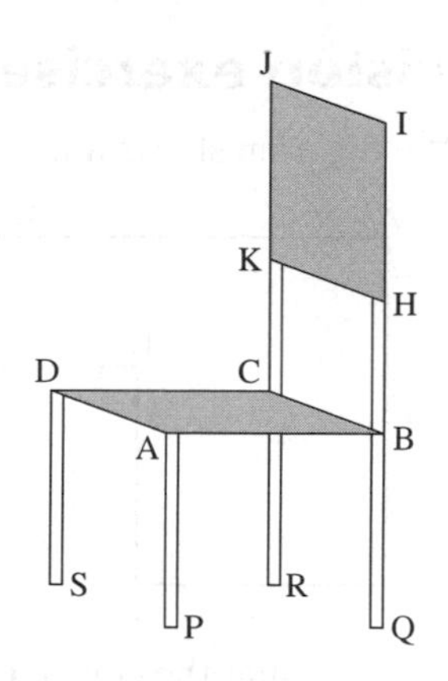

Summary

You should know how to ...	Check out
1 Find the centre of mass of a simple body by symmetry.	**1** A uniform rectangular lamina ABCD has AB = 8 cm and BC = 20 cm. Write down the distance of its centre of mass from the edge AB.
2 Find the centre of mass of a system of particles.	**2** PQRS is a light, rigid, rectangular framework of rods, with PQ = 6 cm and QR = 10 cm. Particles of mass m, $3m$ and $5m$ are placed at P, S and the mid-point of QR respectively. Find the distance of the centre of mass of the system from the edges PQ and QR.
3 Find the centre of mass of a composite body.	**3** The lamina shown comprises a uniform rectangle ABCD and a uniform equilateral triangle CDE of the same density. A 12 cm D 8 cm E B C Find the distance of the centre of mass of the lamina from the edge AB.
4 Find the position of a body when it is suspended in equilibrium.	**4** The lamina in question 3 is suspended in equilibrium from the point A. Calculate the angle that AB makes with the horizontal.

Revision exercise 3

1 The diagram shows a uniform lamina.

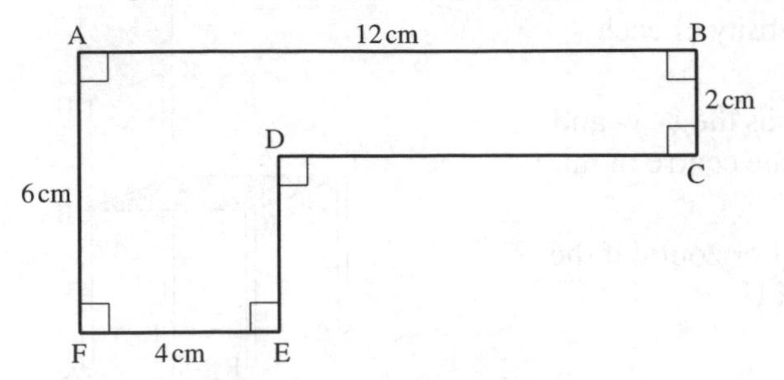

a) Show that the centre of mass of the lamina is 2.2 cm from the side AB.

b) Find the distance of the centre of mass of the lamina from the side AF.

c) The lamina is suspended from the corner A and hangs in equilibrium. Find the angle between the side AF and the vertical.

(AQA, 2002)

M2

2 The diagram shows a uniform lamina.

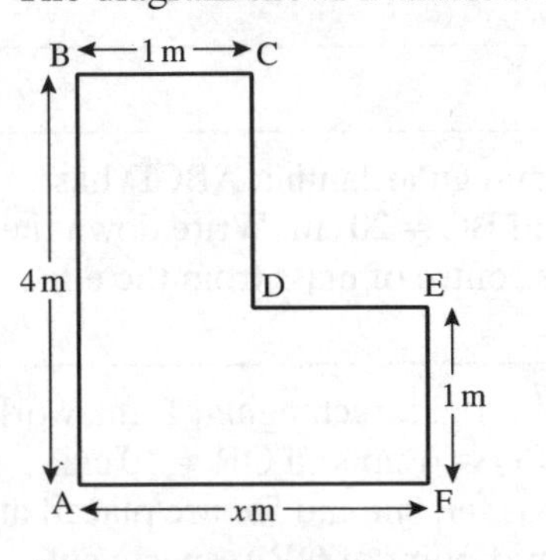

a) For a particular lamina, $x = 7$.

i) Find the distance of the centre of mass of the lamina from the side AB.

ii) The lamina is suspended from the corner C. Find the angle between the side CD and the vertical.

b) Another lamina is suspended from the corner C. Given that the side CD is vertical, find x.

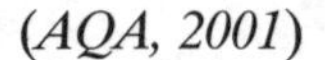

(AQA, 2001)

3 The diagram shows a light square lamina. Particles with masses as shown in the diagram are fixed at the corners and centre of the lamina. The sides of the square are of length 1.2 m.

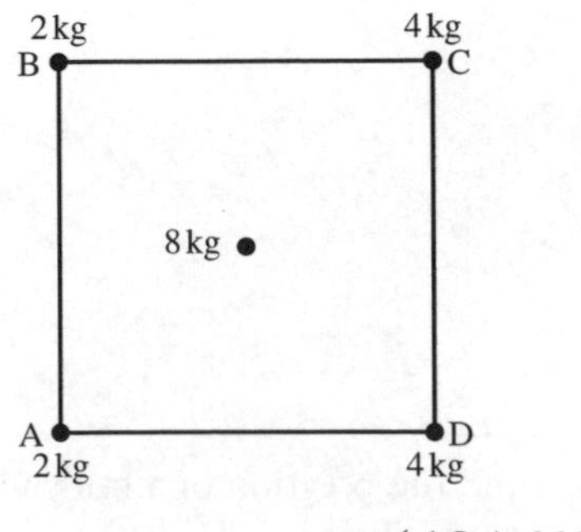

a) Find the distance of the centre of mass of the system of particles from the side AB.

b) Explain why the distance of the centre of mass from the side AD is 0.6 m.

c) The lamina is suspended in equilibrium from the corner A. Find the angle between the side AB and the vertical.

(AQA, 2003)

4 The diagram shows a uniform lamina, which consists of two rectangles ABCD and DPQR.

The dimensions are such that:

DR = PQ = CP = 12 cm;
BC = QR = 8 cm;
AB = AR = 20 cm.

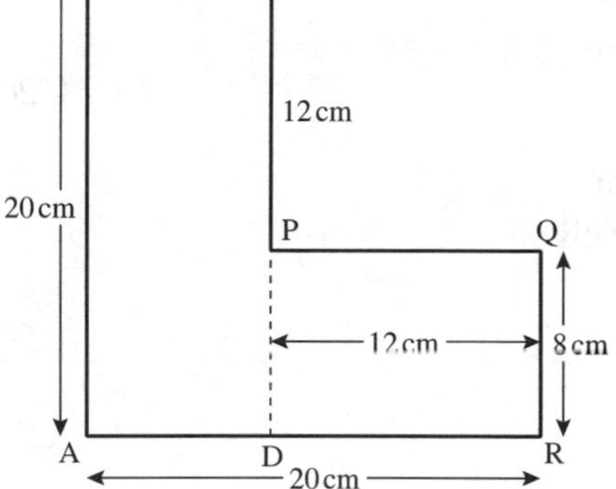

a) Explain why the centre of mass of the lamina must lie on the line AP.

b) Find the distance of the centre of mass of the lamina from AB.

c) The lamina is freely suspended from B.
Find, to the nearest degree, the angle that AB makes with the vertical through B. *(AQA, 2003)*

5 A uniform circular disc is attached to one end of a uniform pole to form the body shown in the diagram.

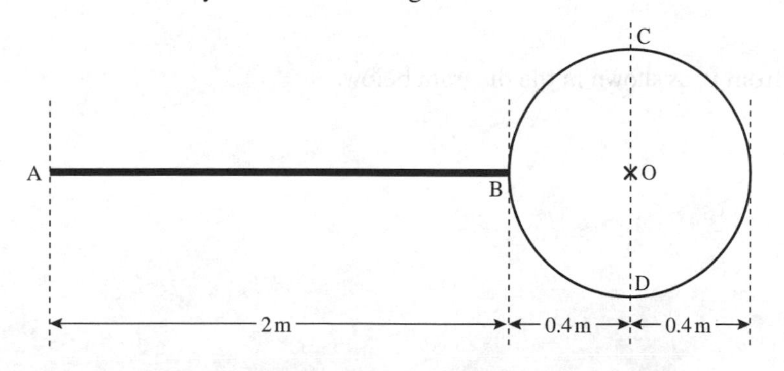

The pole AB has length 2 m and mass 6 kg.
The disc, centre O, has radius 0.4 m and mass 2 kg.
The points A, B and O lie on a straight line.

a) Show that the centre of mass of the body is 1.35 m from A.

b) The diameter, CD, of the disc is perpendicular to OA. The body is suspended from the point C. Find the angle between the pole and the vertical when the body hangs at rest in equilibrium. *(AQA, 2004)*

6 A letter P is formed by bending a uniform steel rod into the shape shown on the right, in which ABCD is a rectangle.

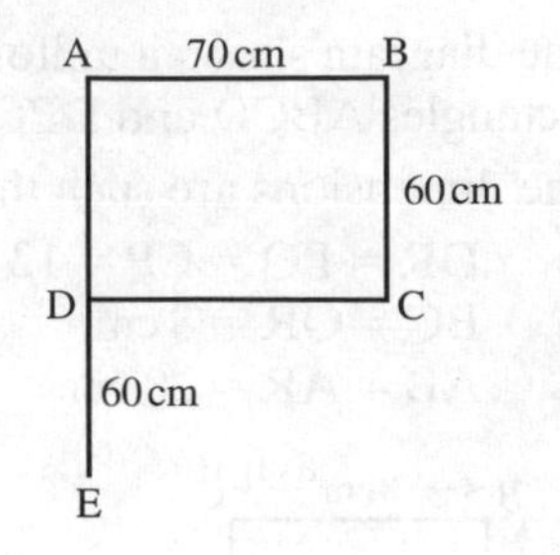

a) Find the distance of the centre of mass of the letter from the side
   i) AE,
   ii) AB.

The letter is to be suspended from a point F on the side AB. The point F is a distance x cm from A.

b) State the value of x if the side AB is to be horizontal.

c) Find the value of x if the side AB is to be at an angle of 5° to the horizontal with A higher than B.

(AQA, 2001)

7 A uniform rectangular plate, PQRS, has mass 1 kg. Particles of mass m kilograms are attached to the plate at Q and R. The plate is shown in the diagram.

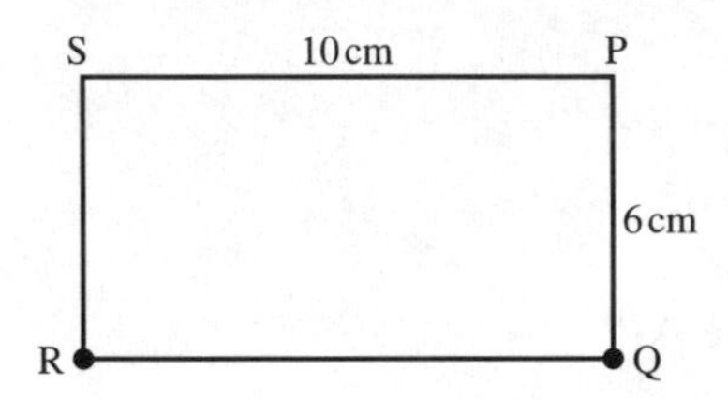

The dimensions of the plate are PQ = SR = 6 cm and PS = QR = 10 cm.

a) State the distance of the centre of mass of the system from PQ.

b) Show that the distance, in centimetres, of the centre of mass of the system from PS is

$$\frac{12m + 3}{2m + 1}.$$

c) The plate is freely suspended from P, as shown in the diagram below.

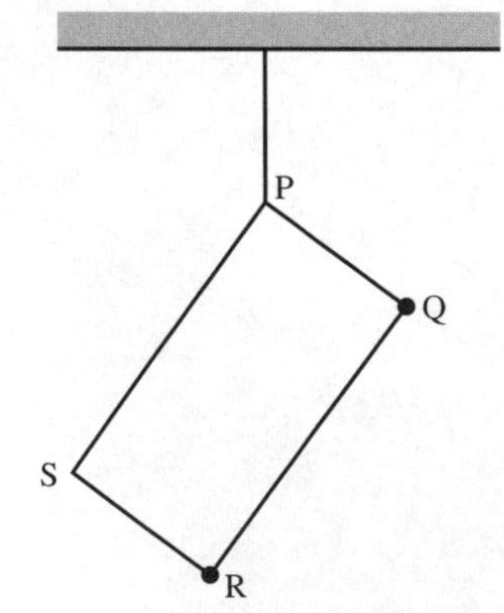

When the plate hangs in equilibrium, PS makes an angle of 45° with the downward vertical. Find the value of m.

(AQA, 2004)

4 Work, energy and power

This chapter will show you how to

- ✦ Calculate the work done by a constant force
- ✦ Calculate gravitational potential energy
- ✦ Calculate kinetic energy
- ✦ Use the principle of conservation of mechanical energy
- ✦ Use the work–energy principle
- ✦ Calculate power
- ✦ Relate power and velocity

Before you start

You should know how to ...	Check in
1 Manipulate algebraic equations.	**1** Find v if $\frac{1}{2}mv^2 + mgh = 3mgh$
2 Manipulate vectors.	**2** Let $\mathbf{r} = 5\mathbf{i} + 7\mathbf{j} - 4\mathbf{k}$, $\mathbf{s} = 3\mathbf{i} - 2\mathbf{j} - 6\mathbf{k}$ Find the magnitude of a) $\mathbf{r} + \mathbf{s}$ b) $5\mathbf{r} - 4\mathbf{s}$
3 Solve quadratic equations by use of the quadratic formula.	**3** Find the roots of $x^2 + 8x - 11 = 0$.
4 Know the equations of motion with constant acceleration.	**4** State the equations of motion for a body moving with constant acceleration.

M2

4.1 Work

If you were to lift a heavy object, drag a packing case along or pedal a cycle, you would know in each case that you were doing work. Work is done whenever a force is applied to alter the motion or position of an object. The amount of work done depends on the magnitude of the force needed and the distance through which the **point of application** of the force moves.

For example, Gladys and Tracy are lifting a 50 kg object using pulleys.

Gladys uses the first arrangement shown. Assume that friction forces in the pulley can be neglected and that the object is being raised at a constant rate.

Resolving vertically for the object, you find

$$T - 50g = 0 \quad \text{(no acceleration)}$$

So, the tension throughout the rope is $50g$ N.

This means that to move the object upwards through a distance of 0.5 m, Gladys has to exert a downward force of $50g$ N on the rope and move it down through a distance of 0.5 m.

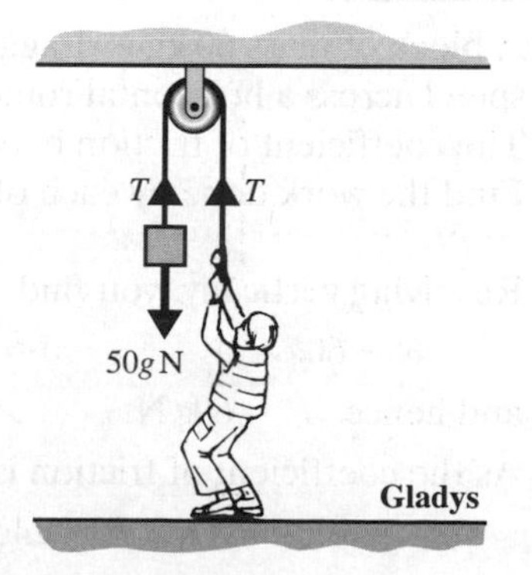

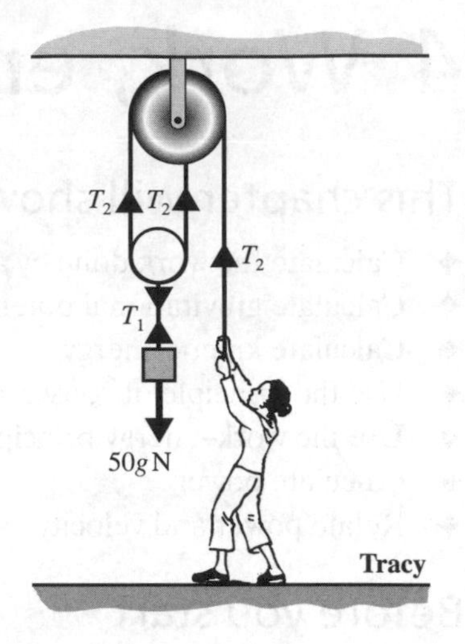

Tracy is not as strong as Gladys, but more sophisticated. She uses the second arrangement shown. Assume that the additional pulley is smooth and has negligible mass.

Resolving vertically for the object, you find

$$T_1 - 50g = 0$$

and hence $T_1 = 50g$ N

Resolving vertically for the small pulley, you find

$$2T_2 - T_1 = 0$$

and hence $T_2 = 25g$ N

This means that to move the object at a constant speed, Tracy has to exert a force of only $25g$ N. However, because the rope is 'shared' between the two sides of the small pulley, to raise the object through 0.5 m, Tracy would have to pull 1 m of rope through the pulley.

M2

Both Gladys and Tracy do the **same amount of work**; that is, they raise a mass of 50 kg through a distance of 0.5 m. Tracy exerts only half the force that Gladys does, but she exerts it through twice the distance to achieve the same effect.

This leads to a definition of work.

> When the point of application of a force F undergoes a displacement s in the direction of the force, the work done is $F \times s$.

The SI unit of work is the **joule (J)**, which is defined as the amount of work done when a force of 1 newton moves a distance of 1 metre.

It should be stressed that the displacement must take place **in the direction of the force**. For example, when a block is dragged along a horizontal surface by a rope, work is done by the tension in the rope and by the friction force. No work is done by the reaction of the surface on the block or by the block's weight, since these forces are perpendicular to the direction of the displacement.

Example 1

A block of mass 60 kg is dragged a distance of 3 m at constant speed across a horizontal rough plane using a horizontal rope. The coefficient of friction between the block and the plane is 0.4. Find the work done by each of the forces acting on the block.

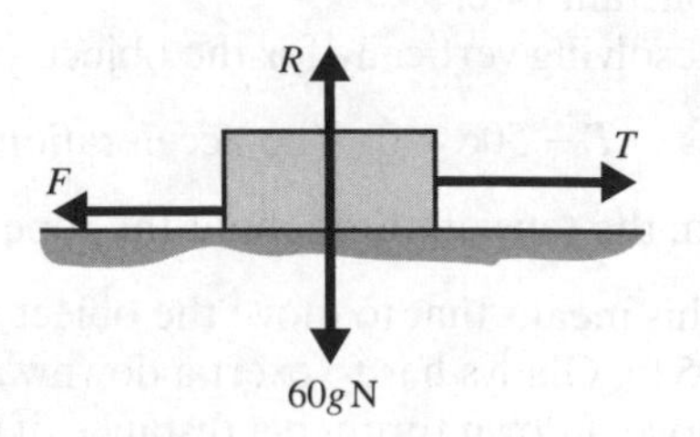

Resolving vertically, you find

$$R - 60g = 0$$

and hence $R = 60g$ N

As the coefficient of friction is 0.4, you have

$$F = 0.4 \times R = 24g \text{ N}$$

Resolving horizontally, you find

$$T - F = 0 \quad \text{(no acceleration)}$$

and hence $T = 24g$ N

There is no vertical displacement of the block, so the work done by the weight and by R is zero.

Displacement in the direction of T is 3 m, so you have

Work done by the tension $= 24g \times 3 = 72g$ J

Displacement in the direction of F is -3 m, so you have

Work done by the friction $= 24g \times (-3) = -72g$ J

Notice the negative work related to friction in Example 1. You can say either that $-72g$ J of work was done by friction, or, commonly, that $72g$ J of work was done **against** friction (that is, work done by the system in overcoming the friction).

Work done against gravity

Suppose an object of mass m is raised at constant speed by a force T.

T

mg

Resolving vertically, you find

$$T - mg = 0$$

and hence $$T = mg$$

Therefore, if the object is raised through a distance h,

Work done by $T = mgh$

Work done by the object's weight $= -mgh$

You say that the

work done against gravity $= mgh$.

M2

Example 2

A plank of length 5 m is inclined so that the higher end is 3 m above the lower. A block of mass 40 kg is towed at constant speed up the whole plank against a friction force of 120 N. Find the total work done by the towing force.

The block undergoes a displacement of -5 m in the direction of the friction force. Therefore

Work done **against** friction $= 120 \times 5 = 600$ J

The block is raised through a height of 3 m. Therefore

Work done **against** gravity $= 40g \times 3 = 1176$ J

Hence the total work done in raising block $= 600 + 1176 = 1776$ J

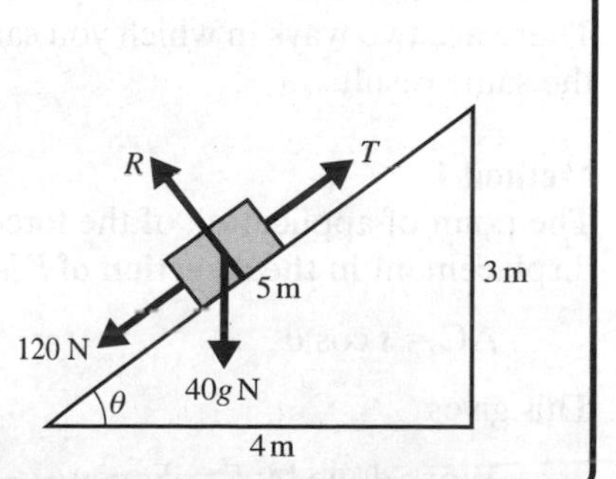

There is an alternative approach to Example 2, as follows.

Resolving parallel to the slope, you find

$$T - 120 - 40g \sin \theta = 0$$

But $\sin \theta = \frac{3}{5}$, so

$$T = 120 + 24g = 355.2 \text{ N}$$

There is a displacement of 5 m in the direction of T, so

$$\text{Work done} = 355.2 \times 5 = 1776 \text{ J}$$

The concept of work done against gravity is particularly useful when the path of the object is not a straight line, as in the following example. Here it would not be helpful to resolve the forces first.

Example 3

An object of mass 8 kg is dragged at constant speed up a surface forming a quarter of a circle of radius 2 m, against a constant frictional resistance of 45 N. Find the total work done.

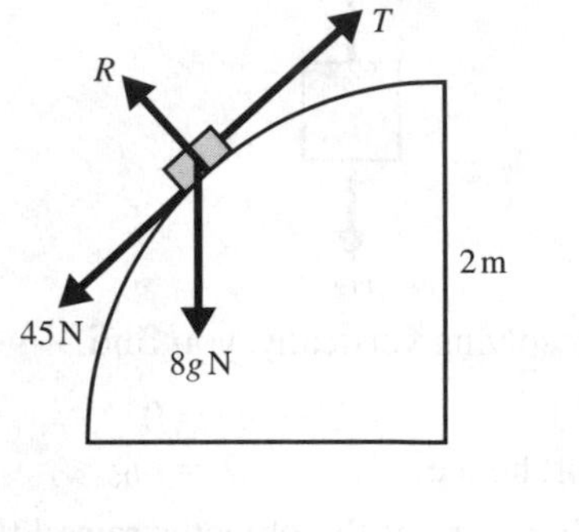

The object is raised through 2 m. Therefore, you have

$$\text{Work done against gravity} = 8g \times 2 = 156.8 \text{ J}$$

The object travels a distance of $-\left(\frac{1}{4} \times 4\pi\right) = -\pi$ m in the direction of the friction force. Therefore, you have

$$\text{Work done against friction} = 45\pi = 141.4 \text{ J}$$

So, you have

$$\text{Total work done} = 156.8 + 141.4 = 298.2 \text{ J}$$

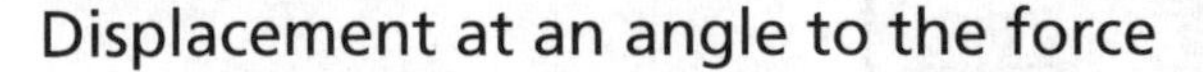

Displacement at an angle to the force

Frequently, the force applied is directed at an angle to the direction in which displacement occurs. For example, if a block is dragged along a horizontal surface using a rope, the rope may be inclined to the horizontal.

Suppose a force F is applied to an object that is then displaced by a distance s in a direction making an angle θ to the direction of F. There are two ways in which you can think about this, each leading to the same result.

Method 1

The point of application of the force moves from A to B, but the displacement **in the direction of** F is represented by AC, where

$$\text{AC} = s \cos \theta$$

This gives

$$\text{Work done by } F = Fs \cos \theta$$

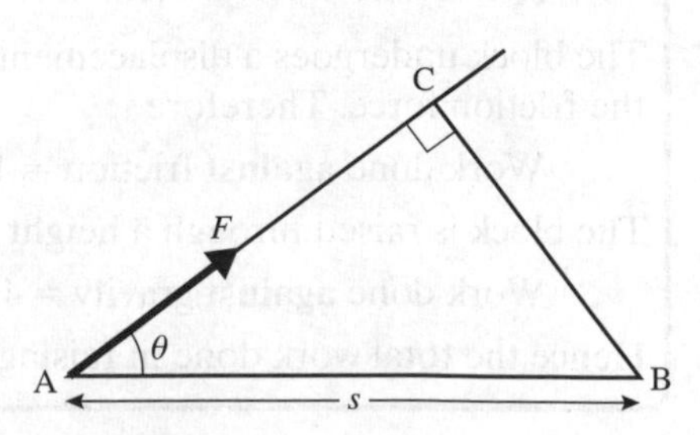

Method 2

The force F can be resolved into two components parallel and perpendicular to the direction of the displacement, as shown.

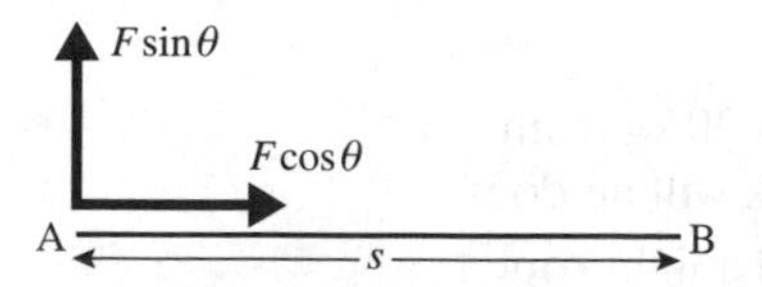

The perpendicular component, $F \sin \theta$, does no work because there is no displacement in that direction.

The parallel component, $F \cos \theta$, is displaced a distance s. Therefore,

Work done by $F = (F \cos \theta) \times s = Fs \cos \theta$

Example 4

A packing case is dragged a distance of 8 m along a horizontal surface by a rope inclined at 40° to the horizontal.
The tension in the rope is 500 N. Find the work done by the tension.

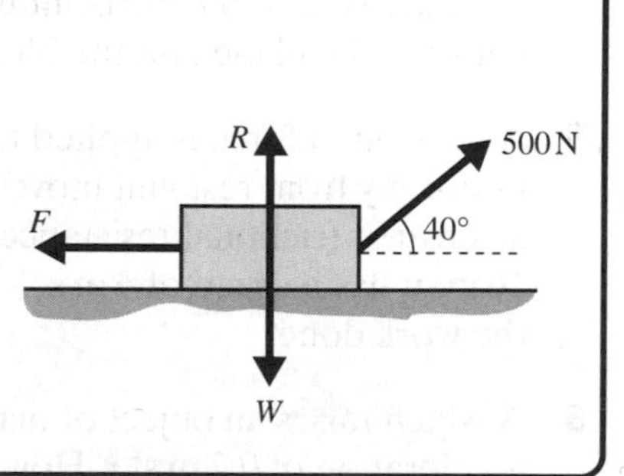

The component of the tension in the direction of motion is

$500 \cos 40°$ N

The displacement in this direction is 8 m.

Therefore, you have

Work done $= 500 \cos 40° \times 8 = 3064.2$ J

M2

Work in vector terms

Note This topic is not on the M2 specification.

Both force **F** and displacement **s** are vectors. So

Work done $= |\mathbf{F}||\mathbf{s}| \cos \theta$

where θ is the angle between the directions of **F** and **s**.

As work is a scalar quantity, if **F** and **s** are given in component form, you can simply find the work done by each of the components and add them together.

Example 5

A force $\mathbf{F} = (5\mathbf{i} + 2\mathbf{j} - \mathbf{k})$ N is applied to an object which undergoes a displacement $\mathbf{s} = (4\mathbf{i} - \mathbf{j} + 6\mathbf{k})$ m.
Find the work done by the force.

In the **i**-direction you have a force of 5 N and a displacement of 4 m.
Hence the work done in this direction is $5 \times 4 = 20$ J.

Similarly the work done in the **j**-direction is $2 \times (-1) = -2$ J
and the work done in the **k**-direction is $(-1) \times 6 = -6$ J.

Hence the total work done is $20 - 2 - 6 = 12$ J.

Exercise 4A

1 Find the work done by a crane which raises a load of 250 kg at a constant speed through a distance of 5.6 m.

2 A man of mass 85 kg wants to move a load of mass 30 kg from ground level on to staging 6 m up. How much work will he do if

a) he hoists it up using a single smooth pulley and a light rope

b) he carries it up a ladder?

3 Blocks, each of mass 5 kg and height 10 cm, are lying side by side on the ground. How much work would be involved in making a stack

a) ten blocks high

b) n blocks high?

4 A block of mass 15 kg is pulled a constant speed for a distance of 12 m across a rough horizontal plane. The coefficient of friction between the plane and the block is 0.6. Find the work done.

5 A horizontal force is applied to a 6 kg body so that it accelerates uniformly from rest and moves across a horizontal plane against a constant frictional resistance of 30 N. After it has travelled 16 m, it has a speed of $4\ \mathrm{m\,s^{-1}}$. Find the applied force and hence the work done.

6 A winch raises an object of mass 20 kg from rest with an acceleration of $0.2\ \mathrm{m\,s^{-2}}$. How much work is done by the winch in the first 12 seconds?

7 Find the work done in pulling a packing case of mass 80 kg a distance of 15 m against a constant resistance of 150 N

a) on a horizontal surface

b) up an incline whose angle θ to the horizontal is given by $\sin\theta = \frac{1}{8}$.

8 A ramp joining a point A to a point C, which is 8 m higher than A, consists of a straight section and an arc of a circle centre B, as shown.

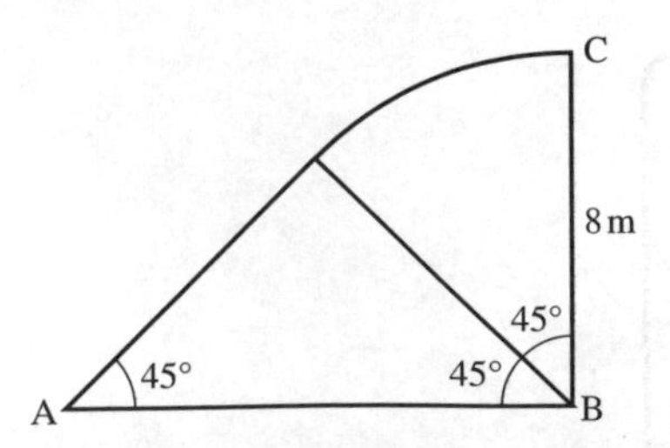

A trolley, which has mass 10 kg and a constant resistance to motion of 140 N, is pulled at a steady speed from A to C. Find the total work done.

9 A body of mass 10 kg is at rest on a rough horizontal surface. The coefficient of friction between the body and the surface is 0.5. A force of 100 N is applied to the body for a period of 20 seconds in one of three different directions, as shown below.

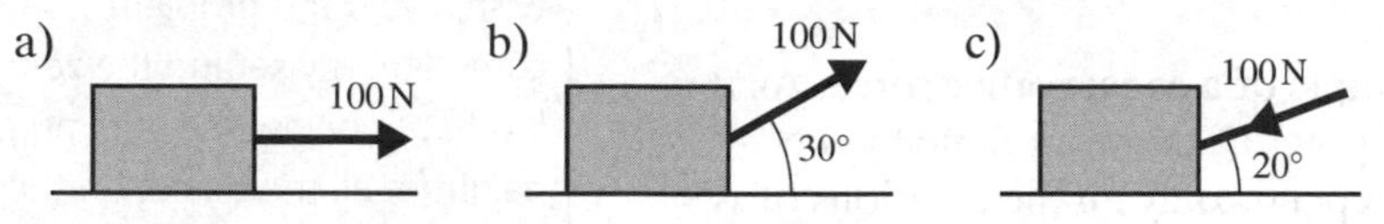

In each case, find the distance travelled by the body and hence the work done by the applied force.

10 Find the work done by a force $\mathbf{F} = (5\mathbf{i} + 3\mathbf{j})$ N whose point of application undergoes a displacement $\mathbf{s} = (3\mathbf{i} + 7\mathbf{j})$ m.

11 A force $\mathbf{F} = (5\mathbf{i} + 4\mathbf{j})$ N acts on a particle which moves from point A, position vector $(\mathbf{i} + 3\mathbf{j})$ m, to point B, position vector $(4\mathbf{i} + 5\mathbf{j})$ m. Find the work done by the force.

12 A force $\mathbf{F} = (2\mathbf{i} - \mathbf{j} + 3\mathbf{k})$ N acts on a particle which moves from point A, position vector $(-2\mathbf{i} + \mathbf{j} - 3\mathbf{k})$ m, to point B, position vector $(5\mathbf{i} - \mathbf{j} + 3\mathbf{k})$m. Find the work done by the force.

M2

4.2 Energy

Gravitational potential energy

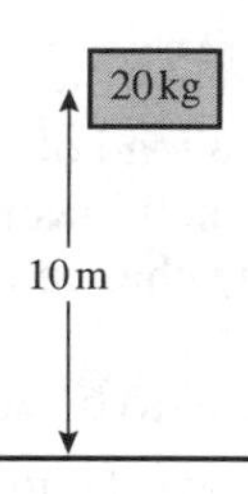

Suppose an object of mass 20 kg is raised through a height of 10 m. The work done against gravity is $20g \times 10 = 1960$ J.

If the object is now allowed to sink back to its original level, its weight will do 1960 J of work.

By raising the object 1960 J of work was 'stored', which was 'retrieved' by lowering the object.

This principle of storing work is used in many ways. For example, some wall-clocks are powered by means of weights suspended from a chain passing over a cog. The heavier weight is raised manually and then, as it gradually descends, it does the work needed to drive the clock. Similarly, some electricity companies use off-peak electricity to pump water from one reservoir into a higher one. At peak times, the water is allowed to run back through turbines, which 'recover' the electricity.

Stored work is called **energy**. In particular, the energy described above, which depends on the position of an object in a gravitational field, is called **gravitational potential energy (GPE)**.

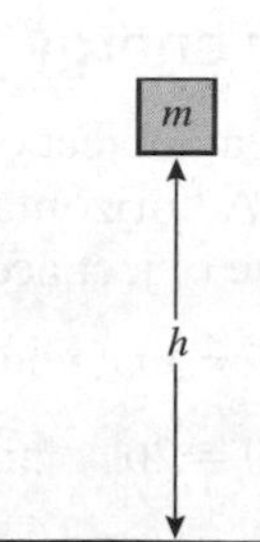

Raising an object increases its GPE, whilst lowering the object decreases its GPE. It is not necessary (or realistic) to talk of the absolute GPE of an object, since the only thing that matters is **change** in its value. It is usual in a given problem to set an arbitrary zero level and measure all GPE in relation to that level.

The work done against gravity in raising an object of mass *m kg* from the zero level to a height of *h* metres is *mgh* J, so

> Gravitational potential energy = *mgh* J

For an object below the zero level, the GPE would take a negative value. You can, if you wish, avoid this problem completely by setting the zero level at or below the minimum height of all the objects in the problem.

The weight of an object is an example of a **conservative** force. As the object moves from a point A to a point B, the work done by the weight (GPE at B − GPE at A) depends only on the positions of A and B and not on the path taken.

If the object follows any closed path, finishing at its starting point, **no work is done by the weight**.

Another example of a conservative force is the tension in an elastic string which is fixed at one end. The work done by the tension when the free end is moved from a point A to a point B depends only on the extension of the string in the two positions. Similarly, the force acting on a metal object in a magnetic field is a conservative force.

See Chapter 5 for more on elasticity and tension.

M2

On the other hand, forces such as friction are not conservative. If friction acts on a body as it moves from a point A to a point B, the work done by friction is greater for a longer path taken. If the body follows a closed path, friction has done a quantity of work dependent on the length of the path. This is called a **dissipative** force.

Example 6

A crane carries a load of 400 kg. It raises it from ground level to a height of 30 m, then lowers it on to a platform 12 m above the ground. Find the change in potential energy in each stage.

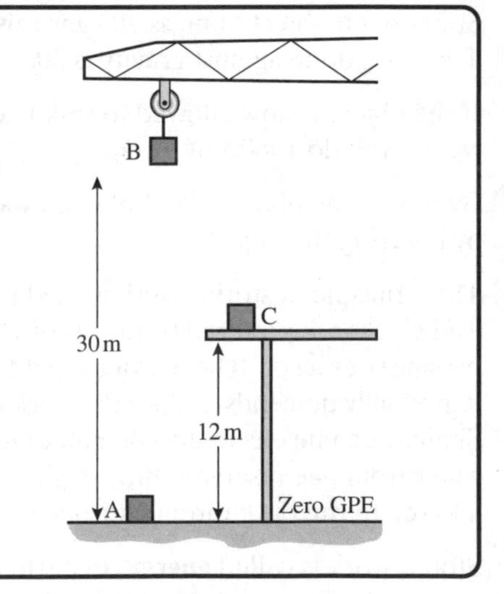

Take ground level to be zero GPE.

The diagram shows the initial (A), intermediate (B) and final (C) positions of the load.

At A: GPE = 0 J
At B: GPE = $400g \times 30 = 117\,600$ J
At C: GPE = $400g \times 12 = 47\,040$ J

During the first stage of motion, GPE increases by 117 600 J.

During the second stage, GPE decreases by $117\,600 - 47\,040 = 70\,560$ J.

Kinetic energy

Suppose an object of mass 20 kg is at rest on a smooth horizontal surface. A horizontal force of 100 N is applied, pulling the object for 10 m. The object accelerates.

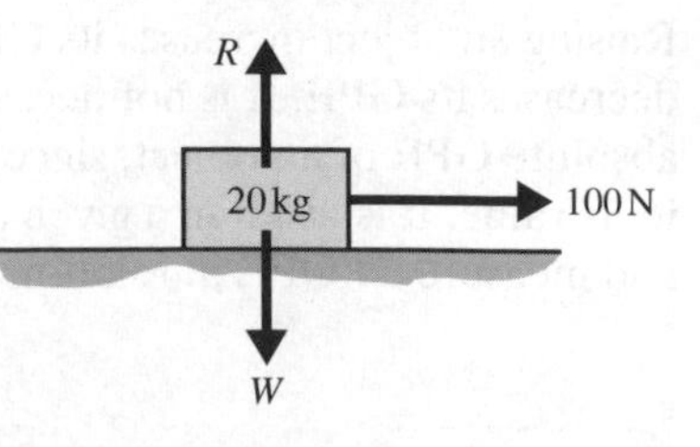

Using $F = ma$, you get

$100 = 20a$ and hence $a = 5\text{ m s}^{-2}$

Using $v^2 = u^2 + 2as$, where $a = 5\ \mathrm{m\,s^{-2}}$, $s = 10$ m and $u = 0$, you get

$$v^2 = 0^2 + 2 \times 5 \times 10 \quad \text{and hence} \quad v = 10\ \mathrm{m\,s^{-1}}$$

The work done in giving the object this speed is 100 N × 10 m = 1000 J.

A moving object has the capacity to do work. You would need to apply a force to stop the 20 kg mass and work would be done against that force. For example, suppose you applied a frictional force of 200 N.

Using $F = ma$, you get

$$-200 = 20a \quad \text{and hence} \quad a = -10\ \mathrm{m\,s^{-2}}$$

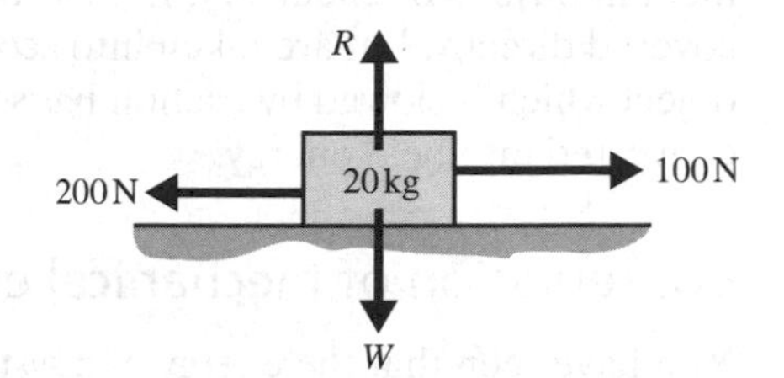

Using $v^2 = u^2 + 2as$, where $a = -10\ \mathrm{m\,s^{-2}}$, $u = 10\ \mathrm{m\,s^{-1}}$ and $v = 0$, you have

$$0^2 = 10^2 - 20s \quad \text{and hence} \quad s = 5\ \mathrm{m}$$

So, the work done against friction is 200 × 5 = 1000 J.

By giving the object a speed of $10\ \mathrm{m\,s^{-1}}$, 1000 J worth of work was 'stored'. This was 'retrieved' when the object was brought to rest.

The work capacity of an object due to its motion is called **kinetic energy (KE)**.

In general, when you apply a force F N to a stationary object of mass m kg for a distance s m, it has acceleration $a\ \mathrm{m\,s^{-2}}$ and final velocity $v\ \mathrm{m\,s^{-1}}$, where

$$v^2 = 2as \quad \text{and hence} \quad as = \tfrac{1}{2}v^2 \qquad [1]$$

Work done = Fs and $F = ma$. Hence

$$\text{Work done} = mas \qquad [2]$$

By substituting from [1] into [2], you get

$$\text{Work done} = \tfrac{1}{2}mv^2\ \mathrm{J}$$

So, an object of mass m kg travelling at $v\ \mathrm{m\,s^{-1}}$ has

> Kinetic energy = $\tfrac{1}{2}mv^2$

Example 7

A particle of mass 4 kg is initially at rest. It is acted upon by a force of 60 N for a period of 8 seconds. Find its kinetic energy at the end of this time.

Using $F = ma$, you find

$$60 = 4a \quad \text{and hence} \quad a = 15\ \mathrm{m\,s^{-2}}$$

Using $v = u + at$, where $a = 15\ \mathrm{m\,s^{-2}}$, $t = 8$ s and $u = 0$, you obtain

$$v = 0 + 15 \times 8 = 120\ \mathrm{m\,s^{-1}}$$

Hence kinetic energy = $\tfrac{1}{2} \times 4 \times 120^2$ = 28 800 J or 28.8 kJ

Alternatively, using $s = ut + \tfrac{1}{2}at^2$ gives $s = 480$ m. Therefore, you have

$$\text{Work done by force} = 60 \times 480 = 28\,800\ \mathrm{J}$$

As all of this work went into accelerating the particle, the kinetic energy is 28 800 J.

Other forms of energy

Gravitational potential energy and kinetic energy are two forms of **mechanical energy**. A third form of mechanical energy – the elastic potential energy of a stretched string or spring – is covered on pages 99–108. There are other, non-mechanical, forms of energy, such as heat, light, electrical, nuclear and chemical energy, which you may meet in physics or chemistry. In this book such forms of energy not covered directly, but are taken into account. For example, a moving object which is slowed by friction has some of its kinetic energy converted into heat energy.

Conservation of mechanical energy

You have seen that the energy of a system is changed if a force does work on it. However, in some situations it is possible to assume that external forces are negligible, and hence that no work is done on the system.

M2

Consider, for example, an object suspended by a string and swinging as a pendulum. Treat it as the usual idealised model: a weightless, inextensible string supporting a point mass whose motion is not subject to air resistance.

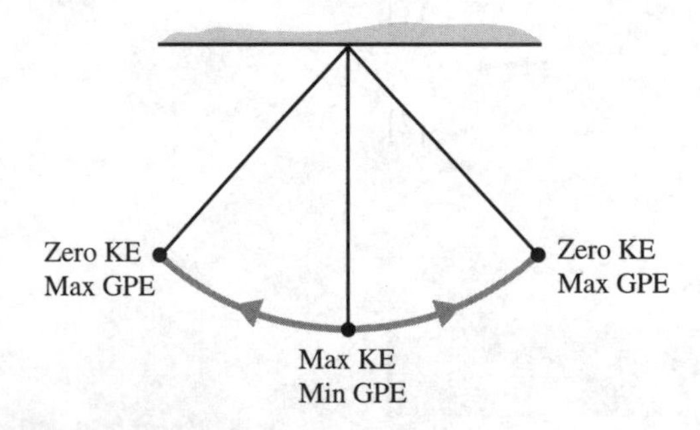

As the particle swings up from its lowest point, work is done against gravity. As a result, the particle slows down – its kinetic energy reduces – but its height, and therefore its gravitational potential energy, increases. This continues until the particle stops, at which point it has zero KE and maximum GPE.

The process then reverses. The height of the particle and therefore its GPE decrease, but its speed and therefore its KE increase. This continues until the particle reaches its lowest position, at which point its GPE is at a minimum and its KE is at a maximum.

In this idealised model, the sequence repeats for ever. On the way up, KE is being converted into GPE. On the way down, GPE is being converted into KE. **The total energy of the particle is exactly the same at all points.**

The total energy of the system is only altered if one or other of the following happens:

- An external force, other than gravity, acts on the system in such a way that work is done.

  When an external force does work on the system, the energy of the system is increased. When the system does work against an external force (for example, friction), the energy of the system is decreased. Mechanical energy can be converted into other forms, such as heat and sound.
- There are sudden changes in the motion of the system.

  This occurs if component particles of the system collide, or if strings connecting component particles are jerked taut. Although the forces involved are internal ones such sudden changes usually involve a loss of energy, such as a conversion to heat, etc.

Gravity has already been allowed for in the GPE.

You can now state the **principle of conservation of mechanical energy**:

> The total mechanical energy of a system remains constant provided no external work is done and there are no sudden changes in the motion of the system.

You will see on pages 100–101 that this principle includes the mechanical energy stored in a stretched elastic string.

It should be stressed that the conditions necessary for mechanical energy to be conserved are never perfectly realised in practice – it is a model based on the usual assumptions. However, the external forces are often sufficiently small to be neglected in the short term, so the principle can be used to make a worthwhile analysis of a situation.

You could, for example, make an excellent prediction of the maximum speed of a pendulum by knowing the height from which it was released. The motion would change over a long time however, due to the small but cumulative effect of air resistance.

Example 8

A particle of mass 2 kg is released from rest and slides down a smooth plane inclined at 30° to the horizontal. Find the speed of the particle after it has travelled 8 m.

Let the start and finish positions of the particle be A and B, and let the speed of the particle at B be v.

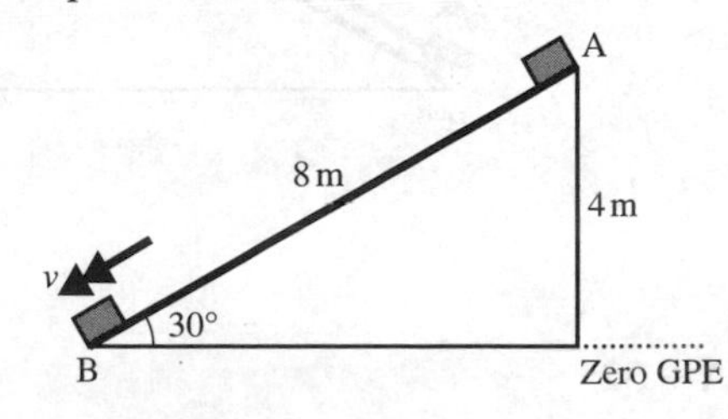

From the diagram, you can see that A is at a height of $8 \sin 30° = 4$ m above B.

At A the energy of the particle is

$KE = 0\,J$
$GPE = 2g \times 4 = 78.4\,J$

and at B its energy is

$KE = \frac{1}{2} \times 2 \times v^2 = v^2\,J$
$GPE = 0\,J$

There are no external forces and no sudden changes, so energy is conserved:

Total energy at B = Total energy at A
which gives $v^2 = 78.4$
and hence $v = 8.85\ m\,s^{-1}$

You could have solved Example 8 using the equations for motion with constant acceleration. Using energy provides an alternative way of approaching the problem and has one major advantage: there is no need to assume that AB is a straight line. This means that if the particle follows a curved path, so that acceleration is no longer constant, you can still use energy considerations to find its final speed.

Work–energy principle

M2

If an external force acts on a system so that work is done, mechanical energy is **not** conserved. You can, however, often make use of energy to solve problems because:

The total work done on the system equals the change in energy.

Example 9 introduces a friction force to the situation in Example 8.

Example 9

A particle of mass 2 kg is released from rest and slides down a plane inclined at 30° to the horizontal. There is a constant resistance force of 4 N. Find the speed of the particle after it has travelled 8 m.

Let the start and finish positions of the particle be A and B, and let the speed of the particle at B be v, as before.
From the diagram, you can see that A is at a height $8 \sin 30° = 4\,m$ above B.

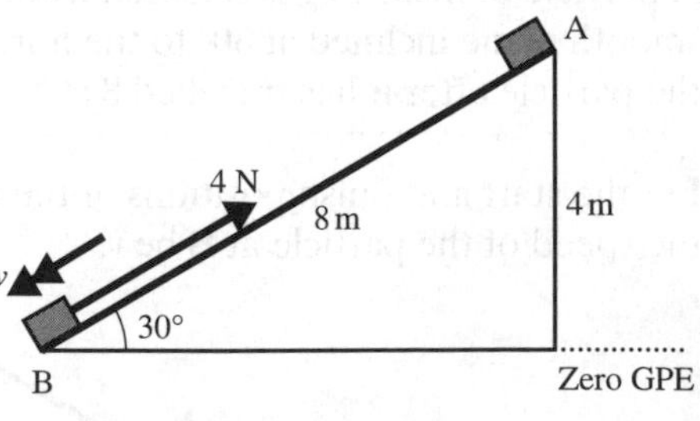

At A the energy of the particle is

$KE = 0\,J$
$GPE = 2g \times 4 = 78.4\,J$

and at B the energy is

$KE = \frac{1}{2} \times 2 \times v^2 = v^2$
$GPE = 0\,J$

The work done against the resistance force is $4 \times 8 = 32\,J$.
Therefore,

Change of energy of system $= -32\,J$
Total energy at B = Total energy at A − 32
which gives $v^2 = 46.4$
and hence $v = 6.81\ m\,s^{-1}$

Example 10

A skateboarder goes down a ramp formed by an arc of a circle of radius 5 m, as shown. She starts from rest at A. The total mass of the skateboader including the board is 50 kg. Find the speed with which she leaves the ramp at B in the following circumstances.

a) There is no appreciable friction.

b) There is a constant resistance force of 10 N.

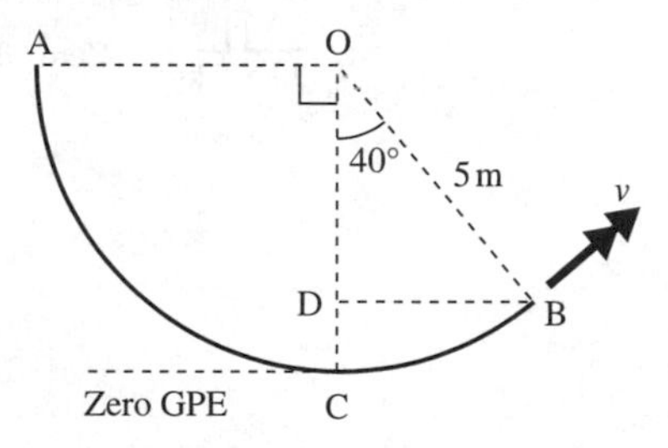

Let the speed at B be v m s^{-1}.

The height of B is CD $= 5 - 5\cos 40° = 1.17$ m.

a) The energy at A is

$$\text{KE} = 0\text{ J}$$
$$\text{GPE} = 50g \times 5 = 2450\text{ J}$$

and at B the energy is

$$\text{KE} = \tfrac{1}{2} \times 50 \times v^2 = 25v^2$$
$$\text{GPE} = 50g \times 1.17 = 573.2\text{ J}$$

Friction is negligible, hence energy is conserved. Therefore, you have

Total energy at B = Total energy at A

which gives $25v^2 + 573.2 = 2450$

and hence $v = 8.66$ m s^{-1}

b) There is a constant resistance force of 10 N, hence work is done on the system. The distance moved by the resistance force is the arc length AB of the ramp.

$$\text{Arc length} = 2\pi \times 5 \times \frac{130}{360} = 11.34\text{ m}$$

Work done against the resistance $= 10 \times 11.34 = 113.4$ J

This is the change in energy of the system.

Therefore, you have

Total energy at B = Total energy at A − 113.4

which gives $25v^2 + 573.2 = 2450 - 113.4$

and hence $v = 8.40$ m s^{-1}

Angle AOB = 130°

Example 11

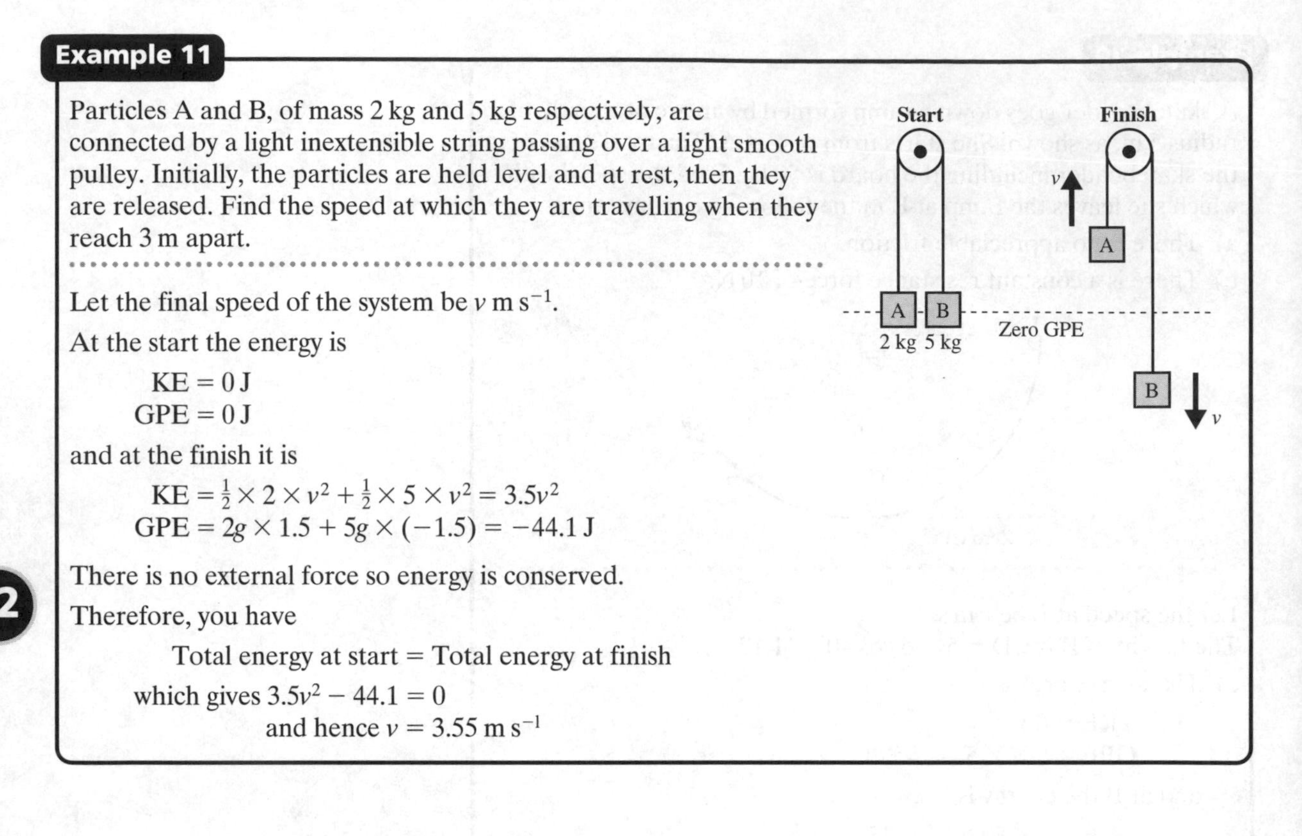

Particles A and B, of mass 2 kg and 5 kg respectively, are connected by a light inextensible string passing over a light smooth pulley. Initially, the particles are held level and at rest, then they are released. Find the speed at which they are travelling when they reach 3 m apart.

Let the final speed of the system be v m s^{-1}.

At the start the energy is

$$\text{KE} = 0\text{ J}$$
$$\text{GPE} = 0\text{ J}$$

and at the finish it is

$$\text{KE} = \tfrac{1}{2} \times 2 \times v^2 + \tfrac{1}{2} \times 5 \times v^2 = 3.5v^2$$
$$\text{GPE} = 2g \times 1.5 + 5g \times (-1.5) = -44.1\text{ J}$$

There is no external force so energy is conserved.

Therefore, you have

Total energy at start = Total energy at finish

which gives $3.5v^2 - 44.1 = 0$

and hence $v = 3.55$ m s^{-1}

M2

Exercise 4B

1 A ball of mass 0.4 kg is thrown vertically into the air at a speed of 25 m s^{-1}. Assuming that air resistance is negligible, use energy methods to find the speed at which the ball is moving when it reaches a height of 20 m. Is the mass of the ball a necessary piece of information?

2 A child of mass 25 kg moves from rest down a slide. The total drop in height is 4 m.

a) Assuming friction is negligible, find the speed of the child at the bottom of the slide.

b) In fact, the child reaches the bottom travelling at 6 m s^{-1}. The length of the slide is 6 m. Find the work done against friction and the average friction force.

3 A particle of mass 2 kg starts from rest at A on a smooth 20° slope. A force of 20 N is applied parallel to the slope, moving the particle up the slope to B, where AB = 3 m. The force then stops acting and the particle continues up the slope, coming instantaneously to rest at C before sliding back down to A. Using energy methods:

a) Find the speed of the particle at B.

b) Find the distance BC.

c) Find the speed of the particle when it returns to A.

4 Particles A and B, of mass 0.5 kg and 1.5 kg respectively, are connected by a light inextensible string of length 2.5 m. Particle A rests on a smooth horizontal table at a distance of 1.5 m from its edge. The string passes over the edge of the table and particle B hangs suspended. The system is held at rest with the string just taut and then released. Find the speed of A when it reaches the edge of the table.

5 Particles A and B, of mass 1 kg and 4 kg respectively, are connected by a light inextensible string of length 4 m. The string passes over a small, light smooth pulley which is 3 m above the ground. The particles are held at rest with A on the ground and B hanging with the string taut. The system is then released.

a) Find the speed of the particles as they pass each other.

b) Explain why you could not have used conservation of energy if the system had started with A on the ground and B held level with the pulley.

M2

6 a) A particle of mass m is attached to the end A of a light rod OA of length a. O is freely hinged to a fixed point and the rod is held in a horizontal position before being released. Find the speed of the particle when the rod makes an angle θ to the downward vertical.

b) If the particle in part a) had been initially projected downwards with speed u, find the value of u for which the rod would just travel round a complete circle.

7 A light rod AB of length $2a$ is freely hinged at O, where OA = a. Particles, of mass m and $2m$ respectively, are attached to the rod at A and B. The rod is held in a horizontal position and then released. Find the maximum speed of B in the subsequent motion.

8 Particles of mass m and $2m$ are connected together by a light inextensible string. Initially, the particles lie at opposite edges of a smooth horizontal table with the string just taut. One of the particles is then nudged over the edge of the table. Find the ratio between the two possible speeds of the system when the other particle reaches the edge of the table.

<u>9</u> A particle is projected with velocity V up a rough plane inclined at an angle θ to the horizontal. The coefficient of friction between the particle and the plane is μ, where μ is sufficiently small so that the particle can slide from rest down the plane. The particle starts at a point A, travels up the plane then slides back down through A. At some point B below A on the plane the particle is again travelling with speed V. Show that

$$AB = \frac{\mu V^2 \cos\theta}{g(\sin^2\theta - \mu^2\cos^2\theta)}$$

4.3 Power

My mass is 90 kg. If I were to climb a flight of stairs taking me to a height of 15 m, I would do $90g \times 15 = 13\,230$ J of work.

This would be the same whether I ran up the stairs or walked up slowly. However, the effect of running up stairs on my breathing and heart rate is quite different from that of walking slowly. The rate at which the work is done is clearly important.

The same thing applies in many situations. For example, you need a more powerful pump to empty a tank in half an hour than to do the same job in five hours.

- ✦ The rate at which work is done is called **power**.
- ✦ The SI unit of power is the **watt (W)**.
- ✦ **1 W** is the rate of working of **1 J s^{-1}**.

If work is done at a variable rate, the relation between work, W, and power, P, can be expressed in calculus terms as $P = \frac{dW}{dt}$.

More simply you can find the average power as

$$\text{Average power} = \frac{\text{work done}}{\text{time taken}}$$

Example 12

A crane lifts a load of 50 kg to a height of 12 m in a time of 20 s. Find the power required.

The work done against gravity is $50g \times 12 = 5880$ J

Assuming that you can neglect any resistance forces, this is the work done by the crane. Therefore, you have

Time taken to lift load = 20 s

Rate of working of crane = $5880 \div 20 = 294$ W

rate = work done ÷ time taken

So, the power required is 294 W.

Example 13

A car of mass 900 kg moves at a steady speed of 15 m s^{-1} up a slope inclined to the horizontal at an angle of $\sin^{-1} 0.2$. Resistance forces total 400 N. Find the power output of the engine.

In 1 second the car travels 15 m on the slope, which raises it through a vertical height of $15 \times 0.2 = 3$ m.

$$\text{Work done against gravity} = 900g \times 3 = 26\,460 \text{ J}$$
$$\text{Work done against resistance} = 400 \times 15 = 6000 \text{ J}$$
$$\text{Total work done in each second} = 32\,460 \text{ J}$$

Hence the rate of working = 32 460 W or 32.46 kW, and this is the power output of the car engine.

work done against gravity = mgh

work against friction = $R \times$ distance

Example 14

A pump raises water from a tank through a height of 3 m and outputs it through a circular nozzle of radius 3 cm at 8 m s^{-1}. Find the rate at which the pump is working. Ignore any resistance forces.

In each second, the pump raises and accelerates a 'cylinder' of water 8 m long and with radius 3 cm. The volume of this water is

$$V = \pi \times 0.03^2 \times 8 = 0.022\,62 \text{ m}^3$$

You can assume that the water has a density of 1000 kg m^{-3}, so you have

$$\text{Mass of water} = 0.0226 \times 1000 = 22.62 \text{ kg}$$

The water is raised through 3 m, so you have

$$\text{GPE given to the water} = 22.62g \times 3 = 665.03 \text{ J}$$

The water is accelerated from rest to 8 m s^{-1}, so you have

$$\text{KE given to the water} = \tfrac{1}{2} \times 22.62 \times 8^2 = 723.8 \text{ J}$$

The total work done by the pump in each second is

$$665 + 723.8 = 1388.8 \text{ J}$$

Therefore, you have

$$\text{Rate of working} = 1390 \text{ W} \quad \text{(to 3s.f.)}$$

M2

Exercise 4C

1 A man raises a load of 20 kg through a height of 6 m using a rope and pulley. Assume the rope and pulley are light and smooth.

a) What would be the man's power output if he completed the task in 30 seconds?

b) If the man's maximum power output is 180 W, what is the shortest time in which he could complete the task?

2 A crate of mass 80 kg is dragged at a steady speed of 4 m s^{-1} up a slope inclined at an angle θ to the horizontal, where $\sin\theta = 0.4$. The total resistance is 250 N. Find the power required.

3 Arnold filled an empty washing-up liquid bottle with water and squirted a horizontal jet at a friend. The diameter of the nozzle was 4 mm and the water emerged at 10 m s^{-1}. Find his rate of working.

Assume the density of water to be 1000 kg m^{-3}.

4 Arnold's friend got her own back using a stirrup pump, which is a device for pumping water from a bucket. The pump raised the water through a height of 80 cm and emitted it as a jet with speed 8 m s^{-1} through a circular nozzle of radius 5 mm. Find her rate of working.

5 A winch has a maximum power output of 500 W. It is dragging a 200 kg crate up a slope inclined at 30° to the horizontal. The coefficient of friction between the crate and the slope is 0.6.

a) Find the work done in dragging the crate a distance a m up the slope.

b) Hence find the maximum speed at which the winch can drag the crate.

M2

6 A force $\mathbf{F} = (4\mathbf{i} + 5\mathbf{j})$ N acts on a particle, moving it along a straight groove from A to B, where A has position vector $(\mathbf{i} - 2\mathbf{j})$ m and B, $(5\mathbf{i} + \mathbf{j})$ m. The process takes 6 seconds. Find the rate at which $\mathbf{F}$ is working.

7 A pump, working at 3 kW, raises water from a tank at $1.2\text{ m}^3\text{ min}^{-1}$ and emits it through a nozzle at 15 m s^{-1}. Find the height through which the water is raised.

8 A horse, which is capable of a power output of 800 W, is able to pull a plough at a constant speed of 1.6 m s^{-1}. Find the resistance to the motion of the plough.

In fact, the traditional unit of power, 1 horsepower, is equivalent to 746 W.

9 A projectile of mass m kg is accelerated at a constant rate up a vertical tube of height h m. When it emerges it rises a further $3h$ m before coming instantaneously to rest. Show that the average rate of working while the projectile is in the tube is $2m\sqrt{6g^3h}$ W.

Relation between power and velocity

Suppose a constant force F N, applied to a particle, just balances the resistance forces, so that the particle has a constant velocity $v\text{ m s}^{-1}$ in the direction of the force.

In 1 second, the particle would travel v m. Therefore, you have

Work done in 1 second $= Fv$ J so

Power exerted by $F = Fv$ W

In situations where the force or the velocity is variable, this expression still gives the power being exerted at a particular instant.

Example 15

A car is being driven at a constant speed of 20 m s^{-1} on a level road against a constant resistance force of 260 N. Find the power output of the engine.

Let the applied force of the engine be F N.

Resolving in the direction of travel and applying Newton's second law, you get

$$F - 260 = 0 \text{ (no acceleration)}$$

and hence $F = 260$ N

Therefore, you have

$$\text{Power} = 260 \times 20 = 5200 \text{ W or } 5.2 \text{ kW}$$

Example 16

A car of mass 800 kg is being driven along a level road against a constant resistance of 450 N. The output of the engine is 7 kW. Find

a) the acceleration when the speed is 10 m s^{-1}

b) the maximum speed of the car.

a) Let the applied force of the engine be F N, and the acceleration be $a \text{ m s}^{-2}$.

Power = Fv, so you have

$$7000 = 10F \quad \text{and hence} \quad F = 700 \text{ N}$$

Resolving in the direction of travel and applying Newton's second law, you get

$$700 - 450 = 800a$$

and hence $a = 0.313 \text{ m s}^{-2}$

b) Resolving in the direction of travel and applying Newton's second law, you get

$$F - 450 = 0 \text{ (no acceleration)}$$

and hence $F = 450$ N

Power = Fv, which gives

$$7000 = 450v$$

and hence $v = 15.56 \text{ m s}^{-1}$

So, the maximum speed is 15.6 m s^{-1}.

As can be seen from Examples 15 and 16, most simple one-dimensional problems involving power can be solved using these two basic equations:

✦ Power = Applied force × Velocity

and

✦ Resultant force = Mass × Acceleration

It is important to stress that in the first equation the force referred to is that exerted by the engine etc, often called the **tractive force**. In the second equation the force is the component in the direction of motion of the resultant of all the forces.

Example 17

A car of mass 900 kg travels up a hill, inclined at 10° to the horizontal, against a constant resistance force of 250 N. Its maximum speed is 45 km h^{-1}. Find

a) the power output of the engine

b) the initial acceleration when it reaches level road at the top of the hill.

....

a) Let the applied force of the engine be F N.
Resolving up the slope and applying Newton's second law, you find

$$F - 250 - 900g \sin 10° = 0 \text{ (no acceleration)}$$

and hence $F = 1781.6$ N

The speed is 45 km h^{-1} = 12.5 m s^{-1}.

Therefore, you have

$$\text{Power} = Fv = 1781.6 \times 12.5 = 22\,269.7 \text{ W}$$
or 22.3 kW (to 3 sf)

45 km h^{-1} R F 250 N 10° 900g N

b) When the car reaches the level, it has the same power and, initially, the same speed. So, F is still 1781.6 N.

Let the initial acceleration be a m s^{-2}.

Resolving horizontally and applying Newton's second law, you find

$$1781.6 - 250 = 900a$$

and hence $a = 1.70$ m s^{-2}

R 45 km h^{-1} 250 N F 900g N

Example 18

A car of mass 1 tonne is towing a trailer of mass 400 kg on a level road. The resistance to motion of the car is 400 N and of the trailer is 300 N. At a certain instant, they are travelling at 10 m s^{-1} and the power output of the engine is 10.5 kW. Find the tension in the coupling between the car and the trailer.

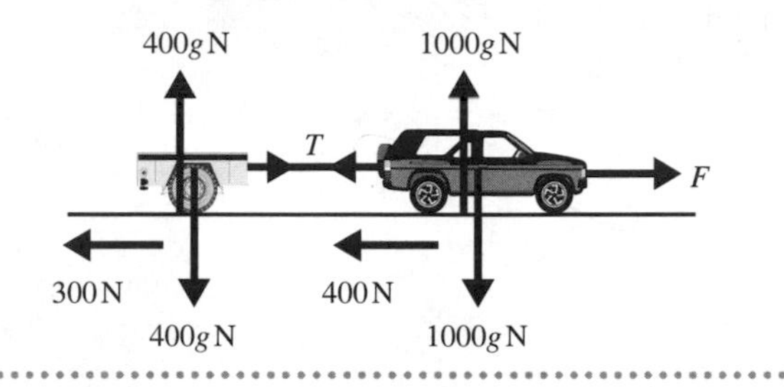

Using Power = Applied force × Velocity, you get

$$10\,500 = 10F$$

and hence $F = 1050$ N

Let the acceleration be a m s^{-2}.

Resolving horizontally for the whole system, you get

$$F - 300 - 400 = 1400\,a$$

and hence $a = 0.25$ m s^{-2}

Resolving horizontally for the trailer only, you get

$$T - 300 = 400 \times 0.25 = 100 \text{ N}$$

So, the tension in the coupling is 400 N.

Problems with variable resistance

So far it has been assumed that resistance to motion is constant. This is never the case in reality, although for slow speeds it may be approximately true. In practice, the resistance is variable and depends on the speed of the vehicle. The nature of the relationship may itself change as the speed increases.

For example, for a small object moving through the air, the air resistance is roughly proportional to the speed v when v is below about 10 m s^{-1}, but for higher speeds (up to about 250 m s^{-1}) the air resistance is proportional to v^2. Around the speed of sound, there are large changes in air resistance and there is no easy relationship with speed, but once the sound barrier is broken the resistance is again roughly proportional to v.

In a practical situation it would be necessary to conduct experiments and decide on a relationship which appeared to fit with the experimental data. This would then be one of the assumptions in the model of the situation.

You can explore the effect of different models using the spreadsheet POWER, which is available on the OUP website at: http://www.oup.co.uk/secondary/mechanics

M2

Example 19

A car of mass 900 kg moves against a resistance that is proportional to its speed. Its power output is 6 kW and on a level road its maximum speed is $40\ \text{m s}^{-1}$. Find its maximum speed up an incline whose angle to the horizontal is θ, where $\sin\theta = \frac{1}{30}$.

Using Power = Applied force × Velocity, you get

$$6000 = 40F$$

and hence $F = 150$ N

Resistance, R, is proportional to speed, v.

So, at maximum speed,

$$R = kv = 40k\ \text{N}$$

Resolving horizontally, you get

$$150 - 40k = 0$$

and hence $k = 3.75$

So, **at any speed**, $R = 3.75v$ N.

Let the maximum speed up the hill be $V\ \text{m s}^{-1}$.

From Power = Applied force × Velocity, you find

$$F = \frac{6000}{V}$$

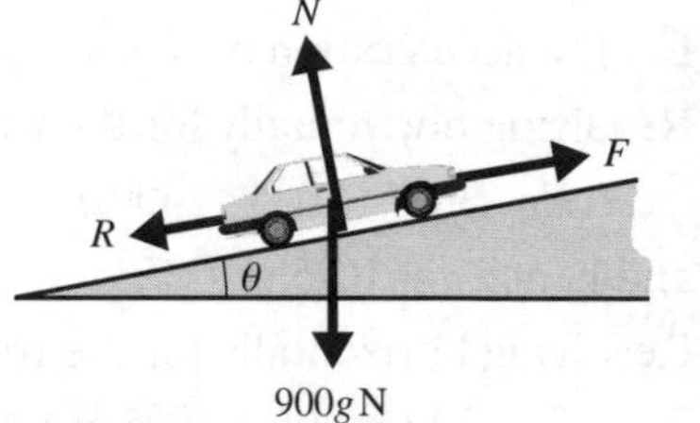

Resolving up the slope, you have

$$\frac{6000}{V} - 3.75V - 900g \times \tfrac{1}{30} = 0$$

Multiplying through by V, you find

$$3.75V^2 + 294V - 6000 = 0$$

and hence $V = 16.8\ \text{m s}^{-1}$ or $-95.2\ \text{m s}^{-1}$

Clearly, $-95.2\ \text{m s}^{-1}$ is inappropriate.

So, the maximum speed up the incline is $16.8\ \text{m s}^{-1}$.

Exercise 4D

1 A train has a maximum speed of $50\ \text{m s}^{-1}$ on the level against a resistance of 40 kN. Find the power output of the engine.

2 A car of mass 800 kg has a maximum speed of $75\ \text{km h}^{-1}$ up a slope against a constant resistance of 500 N. The slope is inclined at $\sin^{-1}\frac{1}{40}$ to the horizontal. Find the power output of the engine.

3 A cyclist of mass 80 kg and power 500 W is travelling on a level road against a constant resistance force of 50 N.
Find a) his maximum speed and b) his acceleration when he is travelling at half his maximum speed.

4 A car of mass 1000 kg has a power output of 5 kW and a maximum speed of 90 km h^{-1} on the level.

a) Find the resistance to motion (assumed to be constant).

b) Find the car's maximum speed up a slope inclined at $\sin^{-1}\frac{1}{25}$ to the horizontal.

5 A train of mass 40 tonnes has a maximum speed of 15 m s^{-1} up a slope against a resistance of 50 kN. The slope is inclined to the horizontal at an angle θ, where $\sin\theta = \frac{1}{50}$. Assuming the resistance is constant, find the maximum speed of the train down the same slope.

6 The frictional resistances acting on a train are $\frac{1}{100}$ of its weight. Its maximum speed up an incline of $\sin^{-1}\frac{1}{80}$ is 48 km h^{-1}. Find its maximum speed on the level.

7 An open truck of mass 5 tonnes is carrying a load of 500 kg of fish up a hill against a constant resistance of 800 N. It is travelling at its maximum speed of 10 m s^{-1}. The hill is inclined at $\sin^{-1}\frac{1}{50}$ to the horizontal. A flock of gulls, mass 1000 kg, descends on the lorry to eat the fish.

a) Find the initial deceleration of the truck and the new maximum speed.

As the truck reaches the top of the hill and moves onto level road, the gulls, having eaten all the fish, fly away.

b) Find the initial acceleration of the truck.

8 A train consists of an engine of mass 50 tonnes and n trucks, each of mass 10 tonnes. The resistance to motion of the engine is 4000 N and that of each truck is 500 N. The maximum speed of the train on the level when there are five trucks is 120 km h^{-1}. Find the power output of the engine and the maximum speed of a train with n trucks going up an incline of $\sin^{-1}\frac{1}{100}$.

9 a) A car of mass 900 kg pulls a trailer of mass 200 kg. The resistance to motion of the car is 200 N and of the trailer is 80 N. Find the power output of the engine if the maximum speed on the level is 40 m s^{-1}.

b) The car and trailer are travelling at 8 m s^{-1} on a hill, inclined at $\sin^{-1}\frac{1}{40}$ to the horizontal. If the resistance is constant and the engine is exerting full power, find the acceleration and the tension in the coupling between the car and the trailer.

10 The resistance to motion of a car is proportional to its speed. A car of mass 1000 kg has a maximum speed of 45 m s^{-1} on the level when its power output is 8 kW. Find its acceleration when it is travelling on the level at 20 m s^{-1} and its engine is working at 6 kW.

11 a) A lorry of mass 10 tonnes has a maximum speed of $20\ \text{m s}^{-1}$ up an incline when its engine is working at 70 kW. The angle of the incline to the horizontal is $\sin^{-1}\frac{1}{100}$. Find the resistance to motion.

b) If the resistance is proportional to the square of the speed, find the maximum speed of the lorry on the level when the engine is working at the same rate.

12 A car has a maximum power P. The resistance to motion is kv. Its maximum speed up a certain slope is V and its maximum speed down the same slope is $2V$. Show that $V = \sqrt{\dfrac{P}{2k}}$.

13 A train of mass 50 tonnes has a maximum speed on the level of $50\ \text{km h}^{-1}$ when the engine is working at 80 kW. Assuming that resistance is constant, how far would the train travel before coming to rest if, when travelling at maximum speed, the engine were disengaged and the train allowed to coast?

M2

14 A cyclist and her cycle have a combined mass of 80 kg. The resistance to motion is proportional to the speed. On the level, she can travel at a maximum speed of $10\ \text{m s}^{-1}$, and she can freewheel down an incline of angle θ at $14\ \text{m s}^{-1}$. Find the maximum speed at which she can go up the same incline.

15 A car working at a rate P W has a maximum speed $V\ \text{m s}^{-1}$ when travelling on the level against a resistance proportional to the square of its speed. At what rate would the car have to work to double its maximum speed?

Summary

You should know how to ...	Check out
1 Calculate the work done by a constant force.	**1** a) A horizontal force of magnitude 400 N acts on a body of mass 5 kg, moving it 20 m in the direction of the force. Calculate the work done by the force. b) Find the work done by the force in part a) if the body moved horizontally but the force was inclined at 25° to the horizontal.
2 Calculate gravitational potential energy.	**2** A body of mass 20 kg falls 4 m vertically. Find the change in its gravitational potential energy.
3 Calculate kinetic energy.	**3** A body of mass 3 kg is moving at 5 m s^{-1}. Find its kinetic energy.
4 Use the principle of conservation of mechanical energy.	**4** A body of mass 8 kg, travelling with speed 3 m s^{-1}, starts to slide down a smooth slope. Find the speed of the body after it has fallen through a vertical displacement of 4 m.
5 Use the work–energy principle.	**5** Suppose the slope in question 4 was rough, giving a constant resistance of 15 N, and in falling 4 m vertically the body travelled 6 m down the slope. What would the final speed of the body now be?
6 Calculate power.	**6** A car travels at a constant speed of 15 m s^{-1} on a level road against a constant frictional force of 180 N. Find the power of the engine.

M2

Revision exercise 4

1 A diver has mass 65 kg. She dives from a fixed diving board, which is 6 m above the level of the water in the pool. When the diver leaves the board, she is travelling vertically upwards and has speed 2 m s^{-1}.

Model the diver as a particle. Assume that there are no resistance forces acting on the diver as she moves through the air and that she does not hit the board on the way down.

a) i) Calculate the kinetic energy of the diver when she leaves the board.
ii) By using an energy method, calculate the maximum height of the diver above the diving board.

b) i) Find the kinetic energy of the diver when she hits the water.
ii) Hence calculate the speed of the diver when she hits the water. *(AQA, 2004)*

2 A ball is projected vertically upwards, from ground level, with an initial speed of 18 m s^{-1}. The ball has a mass of 0.3 kg. Assume that the force of gravity is the only force acting on the ball after it is projected.

a) Calculate the initial kinetic energy of the ball.

b) By using conservation of energy, find the maximum height of the ball above ground level.

c) Find the kinetic energy and the speed of the ball when it is at a height of 2 m above ground level. *(AQA, 2003)*

3 A ball has mass 0.5 kg and is released from rest at a height of 6 m above ground level.

a) Assume that no resistance force acts on the ball as it falls.
   i) Find the kinetic energy of the ball when it has fallen 3 metres.
   ii) Use an energy method to find the speed of the ball when it hits the ground.

b) Assume that a constant resistance force acts on the ball as it falls and that the ball hits the ground travelling at 2 m s^{-1}. Use an energy method to find the magnitude of the resistance force. *(AQA, 2003)*

M2

4 A car, of mass 1000 kg, has a maximum speed of 40 m s^{-1} on a straight horizontal road. When the car travels at a speed $v\text{ m s}^{-1}$, it experiences a resistance force of magnitude $35v$ newtons.

a) Show that the maximum power of the car is 56 000 watts.

b) The car is travelling on a straight horizontal road. Find the maximum possible acceleration of the car when its speed is 20 m s^{-1}.

c) The car starts from rest on a slope inclined at 5° to the horizontal. Find the maximum possible speed of the car as it travels in a straight line up the slope. *(AQA, 2003)*

5 A car, of mass 1200 kg, experiences a resistance force of magnitude $40v$ N when travelling at $v\text{ m s}^{-1}$.

The car travels up a slope inclined at an angle $\sin^{-1}\left(\frac{1}{10}\right)$ to the horizontal. When its speed is 20 m s^{-1} the car is accelerating at 1 m s^{-2}.

a) Show that the power output of the car is 63 520 W.

b) Assume the power calculated in part a) is the maximum for the car. The driver of the car finds that, when travelling up a different slope, the maximum speed of the car is 25 m s^{-1}. Find the angle between this slope and the horizontal. *(AQA, 2002)*

6 A car of mass 1200 kg is being driven up a straight road inclined at 5° to the horizontal. Resistive forces acting on the car total 1960 N.

a) Draw a diagram showing all the forces acting on the car.

b) The car is moving with constant speed 15 m s^{-1}.
   i) Show that the tractive force produced by the engine is approximately 2985 N.
   ii) Determine the rate at which the engine is doing work.

c) The engine has a maximum power output of 60 kW. Find the maximum possible speed of the car up the same slope. *(AQA, 2002)*

7 The diagram shows a children's slide. The curved section AB is one quarter of a circle of radius 2.5 m, so that a child would be travelling horizontally at B. The horizontal section BC has length 5 m. A child of mass 45 kg uses the slide, starting from rest at A.

A

B C

a) A simple model neglects friction and air resistance. Use this model to predict
   i) the kinetic energy of the child at the point C,
   ii) her speed when she reaches the point C.

A revised model assumes that a constant air resistance force of magnitude 20 N acts on the child as she slides from A to C.

b) Calculate the length of the slide between the points A and C. Hence find a revised prediction for the speed of the child at the point C.

A further revision to the model assumes that friction also acts on the child on the section BC and that the coefficient of friction between the child and the slide is 0.5.

c) Find the distance that the child slides beyond the point B. *(AQA, 2001)*

8 A car, of mass 1200 kg, is travelling up a slope at a constant speed of 20 m s^{-1}. The slope is at an angle of 6° to the horizontal. A resistance force of magnitude 420 N also acts on the car when travelling at this speed. In this situation, the power output of the car is a maximum.

a) Show that the maximum power output of the car is 33 000 W to three significant figures.

b) The resistance force acting on the car has magnitude kv newtons, where k is a constant and v m s^{-1} is its speed. Find k.

c) Find the maximum constant speed of the car on a horizontal road. *(AQA, 2004)*

5 Elasticity

This chapter will show you how to

- ✦ Apply Hooke's law to find the tension in elastic strings and springs
- ✦ Find the work done by a variable force
- ✦ Calculate elastic potential energy
- ✦ Apply the principle of conservation of mechanical energy to situations involving elastic strings and springs

Before you start

You should know how to ...	Check in
1 Know the conditions for a rigid body to be in equilibrium.	**1** State two conditions necessary for a rigid body to be in equilibrium.
2 Know the principle of conservation of mechanical energy.	**2** State the principle of conservation of mechanical energy.
3 Know the work–energy principle.	**3** State the work–energy principle.
4 Evaluate definite integrals.	**4** Evaluate $\int_0^e \frac{\lambda x}{l}\,dx$
5 Resolve forces.	**5** Find the component of the force shown in the direction AB. 40 N 35° A B

5.1 Elastic strings and springs

In the problems encountered so far, it has been possible to make the modelling assumption that any strings or rods involved are inextensible. This means that any stretching is negligible compared with the overall length of the string or rod.

This modelling assumption will not always be justified. Many strings, wires and rods can be stretched by significant amounts. It is also necessary to deal with springs, which are designed to stretch and which can also be compressed.

A string or spring that can be stretched, and that regains its original length once the stretching force is removed, is said to be **elastic**. This is to distinguish it from **plastic** items, which can be stretched, but which do not recover when the force is removed.

You know by experience that the amount of stretching produced in a string or spring depends on the magnitude of the force exerted. You should try the following experiments, which are designed to investigate suitable models for the relation between the applied force and the extension produced.

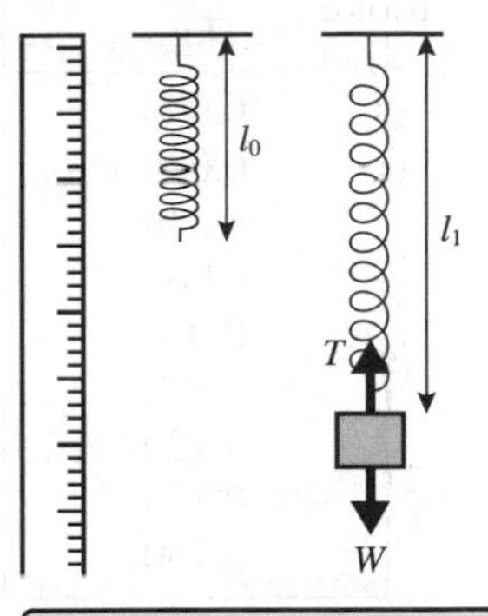

Experiment 1

You need a spring of reasonable length and 'stretchiness', a mass-hanger and masses, and a metre rule or a tape measure.

Fasten the top of the spring to a fixed point and attach the mass-hanger. Measure the unstretched length of the spring. Hang a mass on the spring, and measure the new overall length of the spring. Repeat this for a total of eight different masses.

When the spring is hanging in equilibrium, the tension in it is equal to the downward force supplied by the weight of the attached masses.

Tabulate your results as shown and draw a scatter graph of length against tension. The table and graphs below show the results found by the author.

You can download the spreadsheet ELASTICITY from the OUP website. This will enable you to enter your data in the same format. Just type in: http://www.oup.co.uk/secondary/mechanics

Suspended mass (kg)	Tension (N)	Length (m)
0.000	0.000	0.170
0.020	0.196	0.194
0.040	0.392	0.229
0.060	0.588	0.278
0.080	0.784	0.328
0.100	0.980	0.386
0.120	1.176	0.439
0.140	1.372	0.491
0.160	1.568	0.550

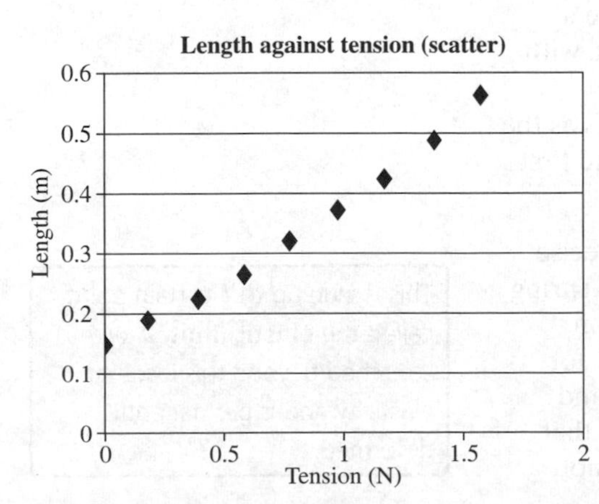

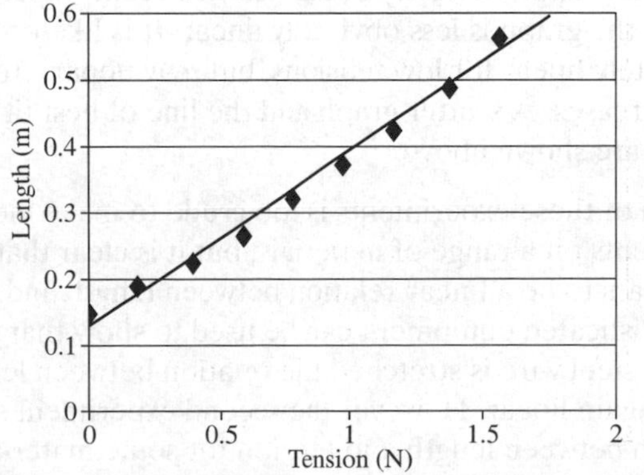

Experiment 2

Repeat Experiment 1, but replace the spring with a piece of elastic, or with a number of elastic bands 'chained' together. The table and graphs shown illustrate the results found by the author.

Suspended mass (kg)	Tension (N)	Length (m)
0.000	0.000	0.265
0.020	0.196	0.285
0.040	0.392	0.312
0.060	0.588	0.347
0.080	0.784	0.390
0.100	0.980	0.446
0.120	1.176	0.510
0.140	1.372	0.585
0.160	1.568	0.665

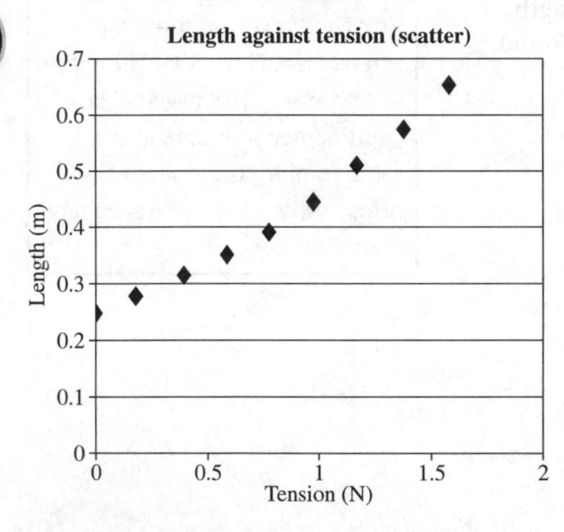

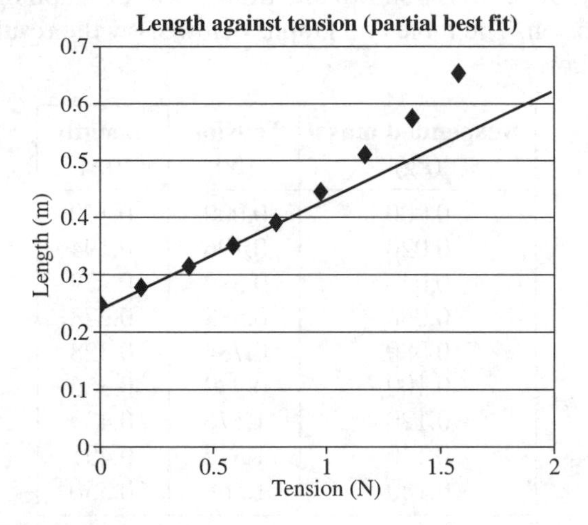

In the first experiment, using the spring, you will probably have a graph which is approximately linear. In the second experiment, with the elastic, the graph is less obviously linear. It is likely to be approximately linear for low tensions, but may depart from this as the tension increases. A scatter graph and the line of best fit for the first five points are shown above.

The design of these experiments is too crude to make more precise measurements on a range of materials, but it is clear that for a spring there appears to be a linear relation between length and tension. More sophisticated equipment can be used to show that when, for example, a steel wire is stretched the relation between length and tension is again linear. However the second experiment shows that the relation between length and tension for some materials is not linear.

This is true up to a certain point, called the **elastic limit**. A wire stretched beyond this loses its elasticity and is permanently deformed.

Interpretation

There appears to be a linear relation between the length of a spring, l, and the tension, T. If the tension is zero, the length is the unstretched, or **natural length**, l_0. The relationship can be written as

$$l = cT + l_0$$

which rearranges to $T = k(l - l_0)$ where $k = \dfrac{1}{c}$.

c and k are both constants.

Notice that $(l - l_0)$ is the extension produced in the spring.

This is the usual model and states that for elastic strings and springs the tension in the string or spring is directly proportional to the extension produced. This model is known as **Hooke's law**, after Robert Hooke, who formulated it in 1678. The model is represented by the equation

$$T = ke$$

where T is the magnitude of the tension
e is the extension of the string or spring from its natural length
k is a constant for a given string or spring.

The constant k is called **stiffness** and depends on the length of the particular string or spring. This is a limitation, because strings or springs that are identical but have different lengths will have different stiffnesses.

A better expression of Hooke's law makes use of the **modulus of elasticity** (λ), which is a constant for all strings of the same material and cross-section, and for all springs of the same construction. In this form, the model is

$$T = \frac{\lambda e}{l}$$

where e is the extension and
l is the natural length of the string or spring.

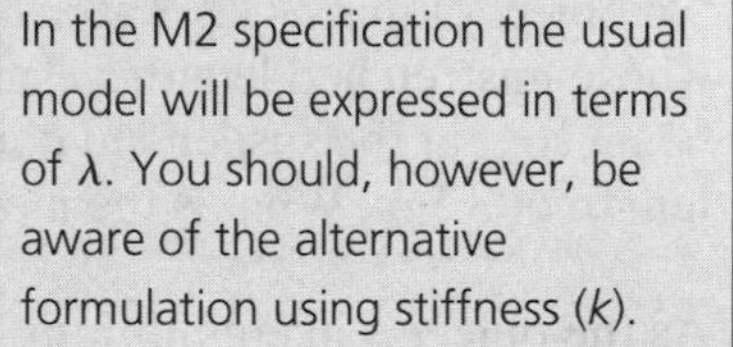
In the M2 specification the usual model will be expressed in terms of λ. You should, however, be aware of the alternative formulation using stiffness (k).

Note that the subscript zero has been dropped in Hooke's law so that the natural length is l rather than l_0.

Modelling assumption

The main assumption in the above model is that the spring or string is **light**. This is clearly reasonable for most strings but a spring probably has a significant mass. This would have had some effect on your results in Experiment 1.

Units

The first version of Hooke's law, $T = ke$, can be arranged as $k = \frac{T}{e}$.

As a result of this, the SI unit for stiffness, k, is the **newton per metre** ($N\,m^{-1}$). The modulus of elasticity, λ, and the stiffness, k, are related by $\lambda = kl$. It follows that the SI unit for λ is the **newton** (N). In fact, λ corresponds to the tension in the string when it is stretched to exactly twice its natural length.

Example 1

An elastic string of length 2 m and modulus of elasticity 49 N has a block of mass 3 kg attached to one end. The other end is fastened to a hook and the block is lowered into its equilibrium position. What is the length of the string in this position?

M2

As the block is in equilibrium, the tension must equal the weight of the block. So, you have

$$T = 3 \times g = 29.4\text{ N}$$

In the Hooke's law equation, $T = \frac{\lambda e}{l}$, you have $\lambda = 49$ and $l = 2$, which gives

$$29.4 = \frac{49e}{2}$$

and hence $e = 1.2\text{ m}$

So, the total length of the string is $2 + 1.2 = 3.2$ m.

Example 2

An object of mass 4 kg is attached to the end of an elastic string whose unstretched length is 3 m. When the string is hung from a beam so that the suspended object is stationary, the string has a length of 3.5 m. What is the modulus of elasticity of the string?

As the object is in equilibrium, the tension in the string must equal the weight of the object. Therefore, you have

$$T = 4 \times 9.8 = 39.2\text{ N}$$

The extension in the string is $e = 3.5 - 3 = 0.5$ m

From $T = \frac{\lambda e}{l}$, you have $39.2 = \frac{0.5\lambda}{3}$

and hence $\lambda = 235.2\text{ N}$

Compressing a spring

Springs are unlike strings as they can be **compressed**. In this case, you have a reduction, e, in the length (a negative extension) and a thrust force (a negative tension) in the spring. It can be shown experimentally that Hooke's law continues to be a good model.

In practice there is a limit to the amount by which a spring may be compressed, as eventually the coils touch. The mathematical model, however, would allow the possibility of compressing the spring until its length was zero or even negative. It is important that you check that the predictions of the model are realistic.

Example 3

A block is fastened to two springs whose natural lengths are 0.5 m and 0.3 m. The springs have modulus of elasticity 15 N and 24 N respectively. The other ends of the springs are fastened to two points, A and C, 0.6 m apart on a smooth table. The block rests on the table so that the springs lie along a straight line.
Find the compressed lengths of the springs.

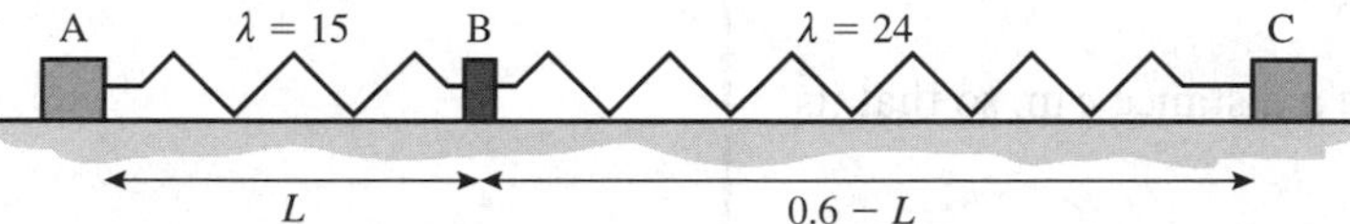

Assume that the thickness of the block is zero.
Let the length AB be L m. The length BC is then $(0.6 - L)$ m.

You then have

	AB	BC
Length	L	$0.6 - L$
Compression	$0.5 - L$	$0.3 - (0.6 - L) = L - 0.3$
Thrust $\left(\text{using } T = \dfrac{\lambda e}{l}\right)$	$\dfrac{15(0.5 - L)}{0.5} = 15 - 30L$	$\dfrac{24(L - 0.3)}{0.3} = 80L - 24$

As the block is in equilibrium, the two thrusts must be equal.
This gives

$$15 - 30L = 80L - 24$$

and hence $L = 0.355$ m

The compressed lengths of the springs are, therefore,
AB = 0.355 m and BC = 0.245 m.

M2

Example 4

A ball of mass 4 kg is fastened between two springs whose natural lengths are 1 m and 0.5 m. The other end of the first spring is fastened to a point, A, and the other end of the second spring to a point B, a distance of 3 m vertically below A. The ball rests in equilibrium. The modulus of elasticity of the springs are 30 N and 10 N respectively. Find the lengths of the springs in this position.

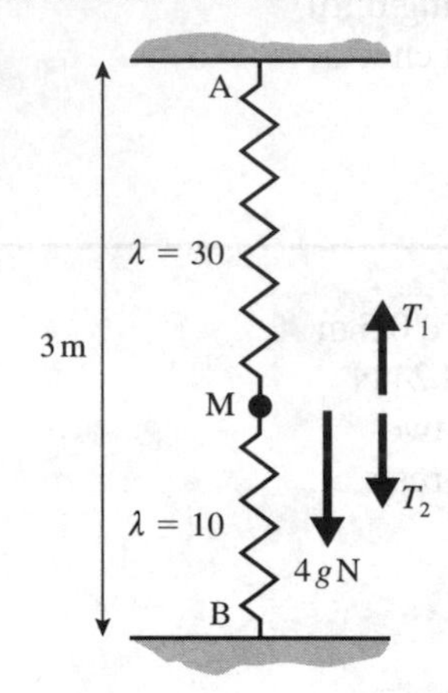

Assume that the spring AM is extended a distance e m, so that its new length is $(1 + e)$ m.

The new length of spring BM is then

$$3 - (1 + e) = (2 - e)\text{ m}$$

The extension of spring BM is then

$$(2 - e) - 0.5 = (1.5 - e)\text{ m}$$

Let the tensions in the springs be T_1 and T_2 as shown in the diagram. By Hooke's law, you have

$$T_1 = \frac{30e}{1} = 30e$$

and

$$T_2 = \frac{10(1.5 - e)}{0.5} = 30 - 20e$$

The ball is in equilibrium. Resolving vertically, you have

$$T_1 - T_2 - 4g = 0$$

and hence $30x - 30 + 20e - 39.2 = 0$

which gives $e = 1.38\text{ m}$

So, the spring AM is 2.38 m long and the spring BM is 0.62 m long.

The next example uses the formulation of Hooke's law in terms of stiffness.

Example 5

You have two elastic strings, each of natural length 1 m and stiffness 32 N m^{-1}, and a ball of mass 4 kg.

a) If the ball is suspended in equilibrium from the ceiling by one of the strings, what is the stretched length of the string?

b) If the two strings are now fastened together end to end and the ball suspended by the combined string, at what distance below the ceiling will it hang at rest?

c) What is the stiffness of the combined string in part b)?

Remember: $T = ke$

a) As the ball is in equilibrium, the tension must equal the weight of the ball. So, you have

$$T = 4g = 39.2\ \text{N}$$

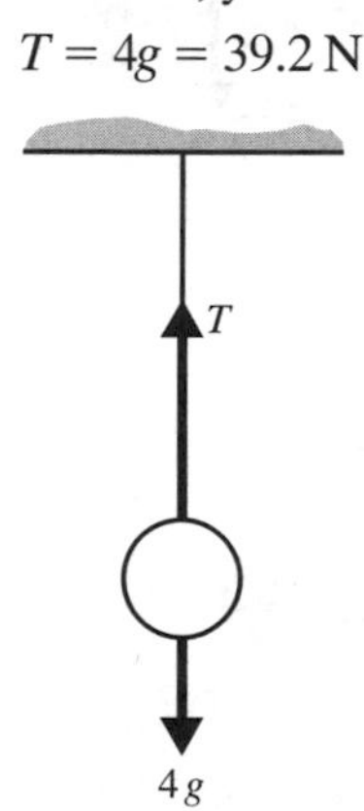

Let e be the extension of the string. From Hooke's law, $T = ke$, you have

$$39.2 = 32e$$

and hence $e = 1.225$ m

The stretched length of the string is therefore 2.225 m.

b) The tension in each of the two strings must equal the weight of the ball. Hence, each string will be stretched by 1.225 m. The distance of the ball below the ceiling is therefore $2 \times 2.225 = 4.45$ m

c) In part b) the total extension is $2 \times 1.225 = 2.45$ m, and the tension is 39.2 N.
Using $T = ke$ you have

$$39.2 = 2.45k$$

and hence

$$k = 16\ \text{N m}^{-1}$$

The stiffness of the combined string is 16 N m^{-1}.

Example 6

Two springs, each of natural length 0.1 m and modulus of elasticity λ N, are fastened at one end to a block, M, of weight 30 N. The other ends are fastened to two hooks, A and B, fixed to the ceiling and 0.16 m apart. The block is lowered until it rests in equilibrium at a distance 0.15 m below the ceiling. Calculate the value of λ.

By symmetry, the tensions in the springs are equal.

You can find the stretched length of the springs by using Pythagoras's theorem on the triangle ACM, which gives

$$AM^2 = 0.08^2 + 0.15^2$$

and hence $AM = 0.17$ m

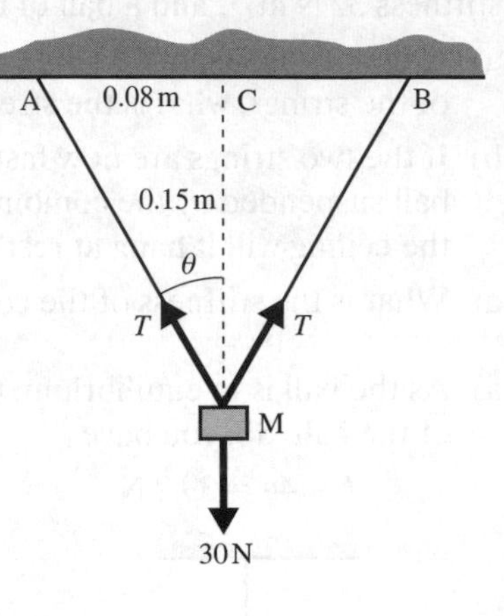

If the springs are inclined at angle θ to the vertical, as shown, you have

$$\cos\theta = \frac{0.15}{0.17} = \frac{15}{17}$$

Resolving vertically, you have

$$2T\cos\theta - 30 = 0$$

which gives $2T \times \frac{15}{17} = 30$

and hence $T = 17$ N

The natural length of each spring is 0.1 m. The stretched length is 0.17 m, so the extension is 0.07 m. The Hooke's law equation $T = \frac{\lambda e}{l}$ then gives

$$17 = \frac{\lambda \times 0.07}{0.1}$$

which gives $\lambda = \frac{17}{0.7} = 24.3$ N

So, the modulus of elasticity of the springs is 24.3 N.

Exercise 5A

1 A block of mass 4 kg is attached to one end of an elastic string whose other end is fixed to the ceiling. The string has a natural length of 2 m and a modulus of elasticity of 90 N. What is the length of the string when the block hangs in its equilibrium position?

2 A spring is designed so that its length doubles when a weight of 50 N is attached to it so that it hangs vertically. When in this position, an extra 20 N is exerted and the extension increases by 10 cm. Find the modulus of elasticity and the natural length of the spring.

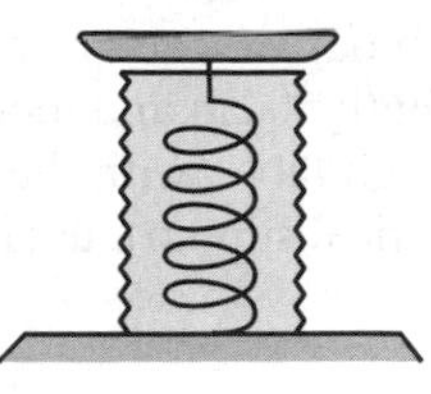

3 A spring is used in a weighing machine, as shown in the diagram. A 2 kg bag of sugar is placed on the scale pan and its height above the table top is measured as 15 cm. When the sugar is replaced by a 1.5 kg bag of flour, the scale pan is 18 cm above the table top.

a) Assuming that the scale pan has negligible weight, what is the height of the scale pan above the table top when there is nothing on the scale pan?

b) What is the modulus of elasticity of the spring?

4 Two springs, the first of natural length 1 m and modulus of elasticity 50 N, the second of natural length 1.5 m and modulus of elasticity 80 N, are fastened side by side to a hook on the ceiling and to a block of mass 5 kg, which hangs at rest below the hook.

a) How far below the ceiling does the block hang?

b) What are the tensions in the springs?

5 Two elastic strings, AB and BC, are joined at B, and the free ends are fixed to points A and C on a smooth horizontal table. The natural length of AB is 85 cm, and of BC is 45 cm. The modulus of elasticity of AB is 45 N and of BC is 65 N. The distance AC is 2.4 m. Find the stretched lengths.

6 A light rod, AB, of length 2 m, is hinged to a vertical wall at A. An elastic string connects B to a point C on the wall, 2 m above A. The string has natural length 1 m and modulus of elasticity 100 N. A block of mass M kg is suspended from B. The system rests in equilibrium. Find the value of M if the angle BAC is

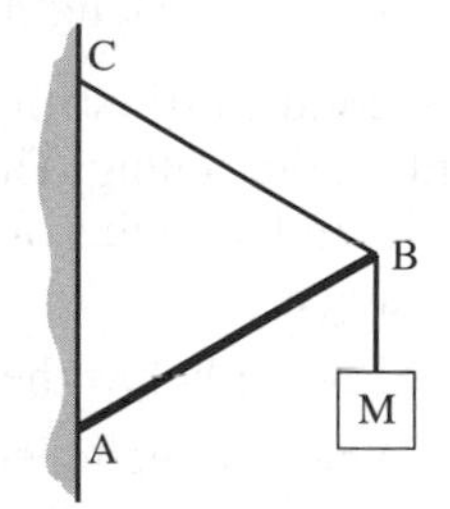

a) 60°

b) 90°

7 A spring of natural length 0.9 m is compressed by a force of 40 N to a length of 0.7 m.

a) What is the stiffness of the spring?

b) If it were stretched by the same force, how long would it be?

8 A spring is stretched by a force of 36 N to a length of 1.2 m. When it is compressed by a force of 24 N, its length is 0.6 m. What are the natural length and stiffness of the spring?

9 A ball of mass 4 kg is attached to an elastic string whose other end is fixed to a point P on a smooth horizontal table. The string has a natural length 1.2 m and modulus of elasticity 28 N. The ball is pulled away from P until it is at a distance of 1.65 m. It is then released. What is the initial acceleration of the ball?

10 A block of wood of mass 2 kg is fastened to a spring of natural length 50 cm and modulus of elasticity 20 N. It is placed on a rough table, with the other end of the spring fastened to a vertical support, as shown.

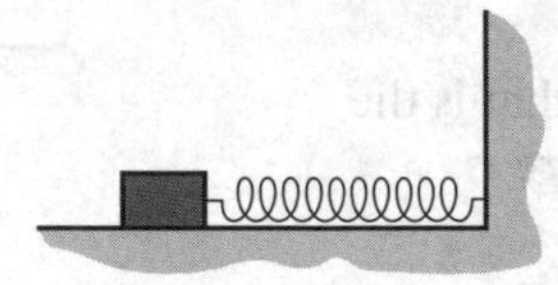

The coefficient of friction between the block and the table is 0.4. If the block is pushed towards the support, what is the closest it can be to the support and remain at rest on the table?

11 A block of mass 2 kg rests on a rough plane which is inclined at 30° to the horizontal. The block is attached to a point at the top of the plane by means of an elastic string of natural length 2 m and modulus of elasticity 100 N. The coefficient of friction between the block and the plane is 0.25. Find the distance between the lowest and highest positions in which the block will rest in equilibrium.

M2

12 A block of mass 2 kg is attached to one end of an elastic string of length 1 m and modulus of elasticity 15 N. The other end of the string is fastened to the ceiling and the block is lowered until it hangs at rest.

a) What is the length of the string in this position?

A second elastic string is attached to the same point on the block and on the ceiling. This second string has a natural length of 0.8 m and modulus of elasticity 8 N. The block is again allowed to hang at rest.

b) How far below the ceiling does the block now hang?

c) What are the tensions in the strings?

13 A block of mass 3 kg rests on a smooth plane which is inclined at 40° to the horizontal. The block is attached to a point at the top of the plane by means of an elastic string of natural length 1.6 m and modulus of elasticity 48 N. Find the stretched length of the string.

14 A ball M of mass 5 kg is attached to the ends of two springs whose other ends are fixed to points A and B, with A 4 m vertically above B. Spring AM has natural length 2 m and modulus of elasticity 100 N. Spring BM has natural length 1 m and modulus of elasticity 20 N. The ball rests in its equilibrium position. What are the lengths of the springs?

15 A block of mass 5 kg is attached to one end of an elastic string of natural length 2 m and modulus of elasticity 80 N. The other end of the string is attached to a fixed point. The block is allowed to hang at rest and is then pulled aside by a horizontal force of 30 N. Find the stretched length of the string.

16 Two identical elastic strings, of stiffness 25 N m^{-1} and natural length 1.5 m, are fastened to a block of mass 2 kg. The other end of one string is fastened to a beam and the other end to the floor, 5 m below the beam.

a) Find the lengths of the strings when the block hangs in equilibrium.

b) If the beam is now gradually lowered, how far above the floor will it be when the lower string becomes slack?

17 An elastic string, of natural length $2l$ and modulus of elasticity λ, is stretched between points A and B, where A is a distance of $4l$ vertically above B. A particle of mass m is attached to the mid-point of the string and is then lowered a distance d until it is in equilibrium. Find d in terms of m, g, l and λ.

18 Springs AB and BC are connected to a block B of mass m. The springs both have stiffness k. The natural length of spring AB is $3l$ and of BC is l. The ends A and C are fastened to points a distance $4l$ apart on a rough horizontal plane. The coefficient of friction between the block and the plane is μ. The block can just rest in limiting equilibrium at the mid-point of AC.

a) Find an expression for k in terms of m, g, l and μ.

b) Find the position of the other point on the line AC at which the block would be in limiting equilibrium.

19 A block, A, of mass m rests on a rough plane inclined at an angle θ to the horizontal. It is on the point of sliding down the slope, and is prevented from doing so by an elastic string, AB, of natural length l and modulus of elasticity λ, which is attached to a point B on the plane above A. The coefficient of friction between the block and the plane is μ.

The point B is then gradually moved up the plane. Show that it can be moved a distance d before the block starts to move up the plane, where

$$d = \frac{2\mu mgl \cos\theta}{\lambda}$$

5.2 Work done by a variable force

Stretching a spring or elastic string, or compressing a spring, requires the application of a force. The point of application of the force moves in the direction of the force during the stretching or compression, and so work is done by the force.

The force, and therefore the work done, varies with the extension. You therefore need to know how to find the work done by a variable force. If a force varies as its point of application moves, it is necessary to use calculus to find the work done.

In the M2 specification the only situations you will meet involve motion along a straight line.

Suppose that force F is a function of the displacement x of its point of application from an origin O. The problem is to find the work done by the force in moving its point of application from $x = a$ to $x = b$.

You know that

Work done = Force × Distance moved in direction of the force

or

$$\mathbf{W} = Fx$$

You can draw the change in force with displacement as a graph. If F is a function of x, the graph could look like this:

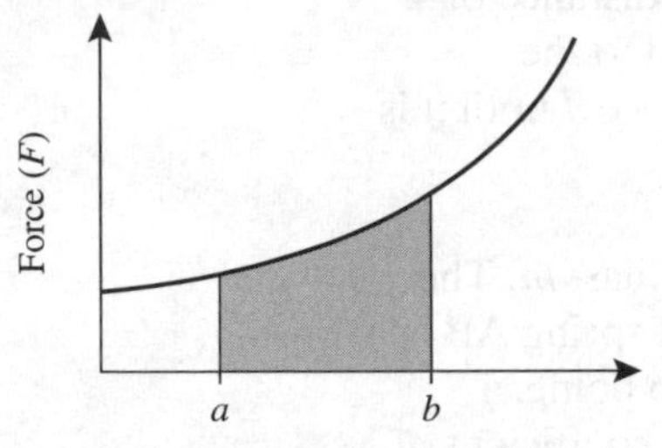

M2

It follows that the work done is equal to the total shaded area under the graph, which can be written as a definite integral.

For a proof of this, see the appendix on page 172.

$$W = \int_a^b F\,\mathrm{d}x$$

Tension, work and energy

In the Hooke's law model, the variable force is the tension $T = \dfrac{\lambda e}{l}$, which gives

$$W = \int \frac{\lambda e}{l}\,\mathrm{d}e = \frac{\lambda e^2}{2l} + c$$

When the extension, e, is zero, no work has been done by the force, so $c = 0$. So, the work done in stretching a string or spring is given by

$$W = \frac{\lambda e^2}{2l} \quad \text{or} \quad W = \tfrac{1}{2}ke^2$$

Although the above refers to stretching a spring, the same equation gives the work done in compressing a spring by an amount e from its natural length.

The tension or compression is a conservative force. The work it does depends only on the initial and final extensions, and is zero if the initial and final extensions are the same. You can say that a stretched string or spring, or a compressed spring, has stored or potential energy. This energy is converted to the form of kinetic energy when the string or spring is released.

For example, suppose you attach one end of an elastic string to an object on a smooth table and the other end to a fixed point on the table. You then pull the string into a stretched position and let go. The object will start to move as the string returns to its original length. The stored energy in the string is being converted into the kinetic energy of the moving object.

This stored energy is equal to the work done in stretching or compressing the string or spring. It is called **elastic potential energy** (EPE), and is the third form of mechanical energy.

$$\text{Elastic potential energy} = \frac{\lambda e^2}{2l} \quad \text{or} \quad \tfrac{1}{2}ke^2$$

The mechanical energy conservation equation now looks like

$$\text{EPE} + \text{GPE} + \text{KE} = \text{constant}$$

or

$$\frac{\lambda e^2}{2l} + mgh + \frac{1}{2}mv^2 = \text{constant}$$

See pages 67–69 for gravitational potential energy (GPE) and kinetic energy (KE).

M2

Example 7

A block of mass 4 kg is attached to one end of a spring, whose natural length is 2 m and whose modulus of elasticity is 128 N. The other end of the spring is fixed to a point, O, on a smooth table. The block is held 3 m from O and then released. Find the speed of the block when the spring reaches its natural length, and investigate the subsequent motion of the block.

O 2 m L 1 m B

Assuming the table top is the zero level for GPE, you have GPE = 0 throughout the motion.

At B, you have

$$\text{EPE} = \frac{\lambda e^2}{2l} = \frac{128 \times 1^2}{2 \times 2} = 32\text{ J}$$

$$\text{KE} = 0\text{ J}$$

So the total energy at B is 32 J.

Let the speed of the block at L be v. Then at L you have

$$\text{EPE} = 0\ \text{J}$$
$$\text{KE} = \tfrac{1}{2}mv^2 = \tfrac{1}{2} \times 4 \times v^2 = 2v^2$$

So, the total energy at L is $2v^2$ J.

There are no outside forces or sudden changes, so energy is conserved. Therefore, you have

$$2v^2 = 32$$

and so $v = 4\ \text{m s}^{-1}$

So, the speed of the block at L is $4\ \text{m s}^{-1}$.

Now consider the subsequent motion. The block passes through L and the spring begins to compress. This slows the block, as kinetic energy is converted to elastic potential energy. The block will become stationary when the EPE is again 32 J, which occurs when the spring is **compressed** by 1 m.

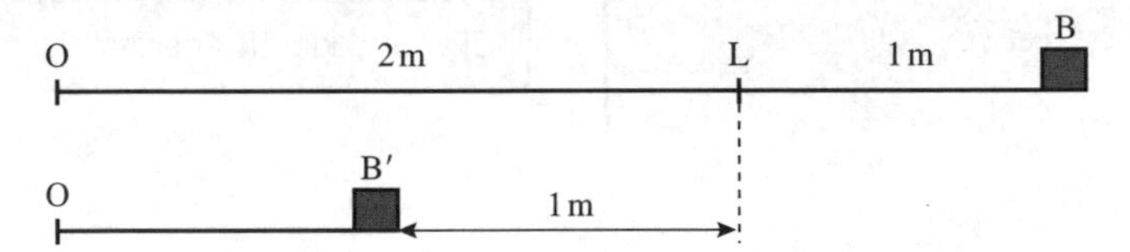

The block would oscillate backwards and forwards about L.

This assumes that the spring is designed to allow this degree of compression without the coils meeting. If they met, there would be an impulse on the block, and the principle of conservation of energy would cease to apply.

The situation in Example 7 would have been different if the spring had been replaced by an elastic string of equivalent length and stiffness. The string would go slack when the block reached L. The block would continue to move through L and O at $4\ \text{m s}^{-1}$.

Eventually, the string would become taut in the opposite direction. The block would slow and become stationary when the string was again extended by 1 m.

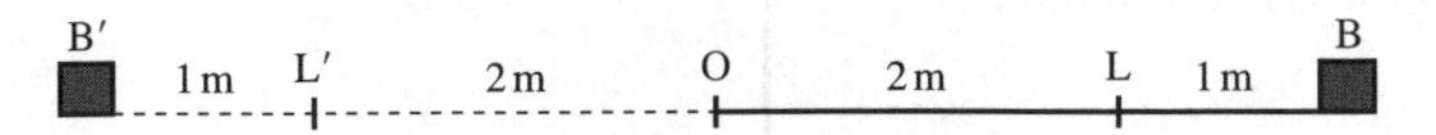

The position of the block would then be at -3 m. The block would then oscillate backwards and forwards about O.

Example 8

A ball of mass 2 kg is attached to one end of an elastic string, which is vertical with its other end fixed to the ceiling. The string has a natural length of 2 m and a modulus of elasticity of 100 N. The ball is held so that the string is at its natural length and is then released from rest. Stating any assumptions made, find the distance the ball drops before coming instantaneously to rest.

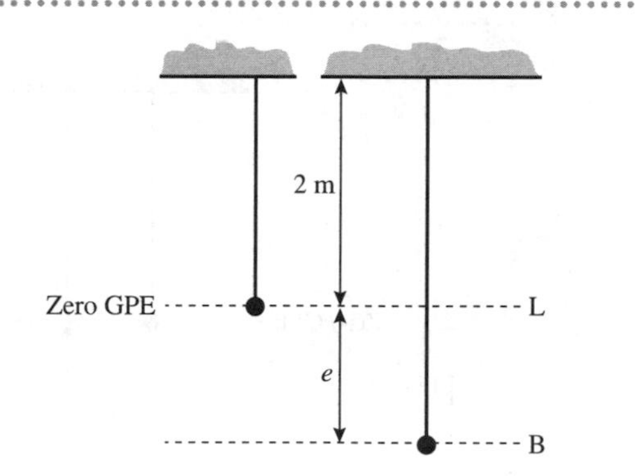

Assumptions:

- The ball is a particle.
- The string is light.
- The string does not deform as it stretches, so that you can reasonably assume Hooke's law is a suitable model throughout the motion.
- There is no air resistance.

At L, you have

KE = 0 since the ball starts from rest
GPE = 0 chosen zero level
EPE = 0 since the string is unstretched

So, the total energy at L is 0.

At B, the lowest position of the ball, you have

KE = 0 since the ball is again at rest

$$\text{GPE} = -mge = -19.6e$$

$$\text{EPE} = \frac{\lambda e^2}{2l} = \frac{100e^2}{2 \times 2} = 25e^2$$

So, the total energy at B is $25e^2 - 19.6e$.

Energy is conserved. So, you have

$$25e^2 - 19.6e = 0$$

and so $\quad e = 0$ or $e = 0.784$ m

This shows that the ball is at rest at its starting point L (when $e = 0$) and again at B when it has dropped a distance of 0.784 m.

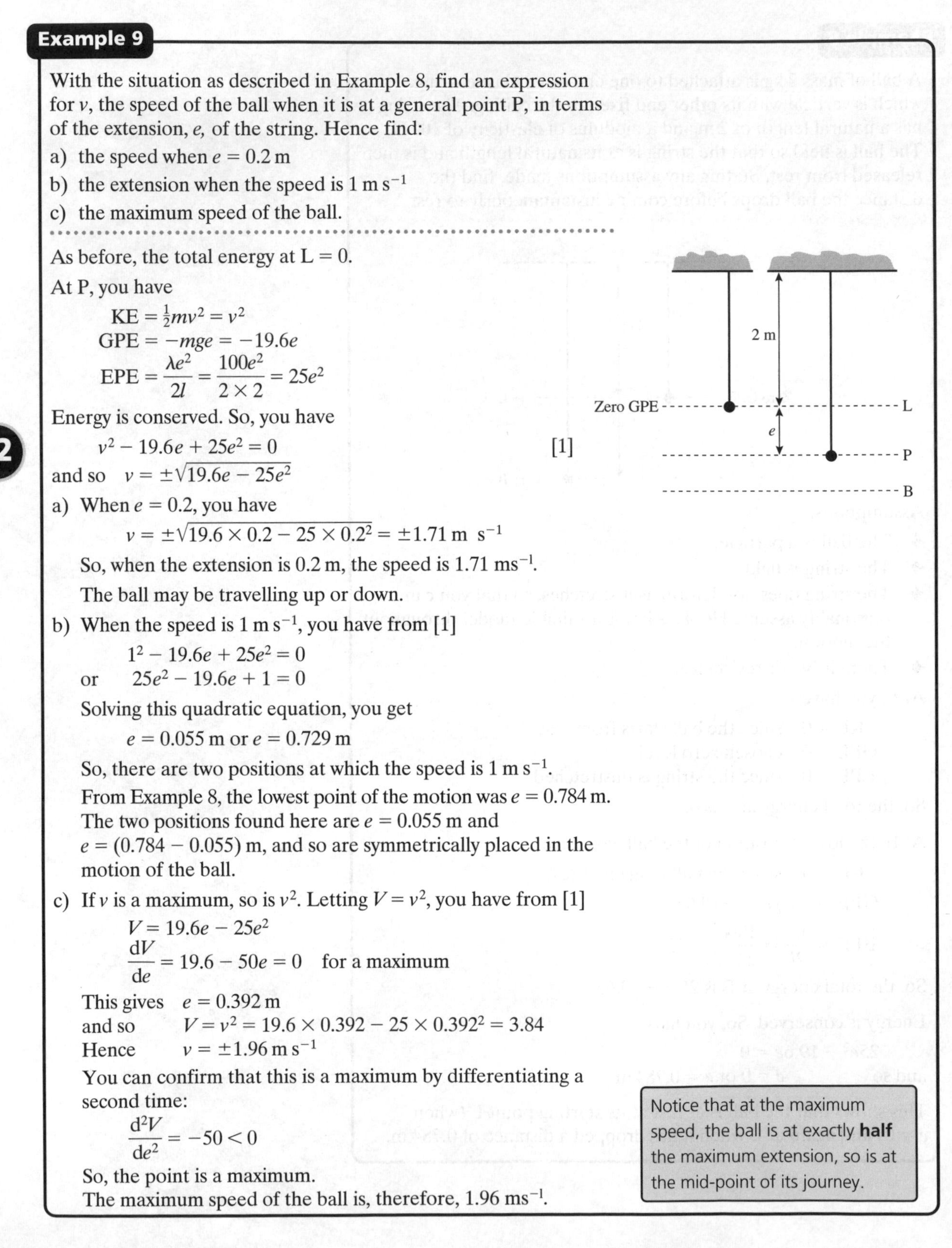

Example 9

With the situation as described in Example 8, find an expression for v, the speed of the ball when it is at a general point P, in terms of the extension, e, of the string. Hence find:

a) the speed when $e = 0.2$ m

b) the extension when the speed is 1 m s^{-1}

c) the maximum speed of the ball.

As before, the total energy at L = 0.

At P, you have

$$\text{KE} = \tfrac{1}{2}mv^2 = v^2$$
$$\text{GPE} = -mge = -19.6e$$
$$\text{EPE} = \frac{\lambda e^2}{2l} = \frac{100e^2}{2 \times 2} = 25e^2$$

Energy is conserved. So, you have

$$v^2 - 19.6e + 25e^2 = 0 \qquad [1]$$

and so $v = \pm\sqrt{19.6e - 25e^2}$

a) When $e = 0.2$, you have

$$v = \pm\sqrt{19.6 \times 0.2 - 25 \times 0.2^2} = \pm 1.71 \text{ m s}^{-1}$$

So, when the extension is 0.2 m, the speed is 1.71 ms^{-1}.

The ball may be travelling up or down.

b) When the speed is 1 m s^{-1}, you have from [1]

$$1^2 - 19.6e + 25e^2 = 0$$

or $25e^2 - 19.6e + 1 = 0$

Solving this quadratic equation, you get

$$e = 0.055 \text{ m or } e = 0.729 \text{ m}$$

So, there are two positions at which the speed is 1 m s^{-1}.

From Example 8, the lowest point of the motion was $e = 0.784$ m.

The two positions found here are $e = 0.055$ m and $e = (0.784 - 0.055)$ m, and so are symmetrically placed in the motion of the ball.

c) If v is a maximum, so is v^2. Letting $V = v^2$, you have from [1]

$$V = 19.6e - 25e^2$$
$$\frac{dV}{de} = 19.6 - 50e = 0 \quad \text{for a maximum}$$

This gives $e = 0.392$ m

and so $V = v^2 = 19.6 \times 0.392 - 25 \times 0.392^2 = 3.84$

Hence $v = \pm 1.96 \text{ m s}^{-1}$

You can confirm that this is a maximum by differentiating a second time:

$$\frac{d^2V}{de^2} = -50 < 0$$

So, the point is a maximum.

The maximum speed of the ball is, therefore, 1.96 ms^{-1}.

Notice that at the maximum speed, the ball is at exactly **half** the maximum extension, so is at the mid-point of its journey.

M2

This is a harder example that involves the addition of a weight. You can still use the principle of conservation of energy to solve the problem.

Example 10

A scale pan of mass 50 g is hanging in equilibrium on an elastic string of natural length 60 cm and modulus of elasticity 10 N. An object of mass 200 g is gently placed on the pan. How far does the pan drop before coming instantaneously to rest?

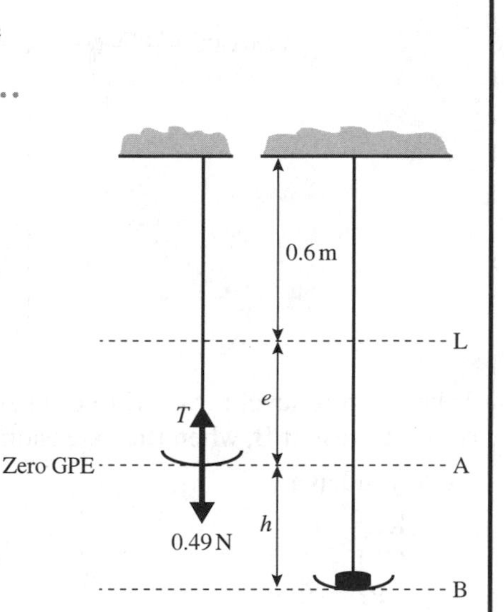

In the diagram, the pan hangs in equilibrium at A, when its extension is e. Its lowest point after the object is placed on it is B, when it has dropped a distance h.

First, you need to find e.

Resolving vertically when the pan is in equilibrium, you have

$$T = 0.05g = 0.49 \text{ N}$$

Using $T = \dfrac{\lambda e}{l}$ you have

$$0.49 = \frac{10e}{0.6}$$

and so $e = 0.0294$ m

The object is now placed on the pan, and you use conservation of energy to find how far it falls.

Take the zero level for GPE at A, as shown.

At A, you have

$$\text{KE} = 0$$

$$\text{GPE} = 0$$

$$\text{EPE} = \frac{\lambda e^2}{2l} = \frac{10 \times 0.0294^2}{2 \times 0.6} = 0.0072 \text{ J}$$

At B, you have

$$\text{KE} = 0$$

$$\text{GPE} = -mgh = -0.25 \times 9.8 \times h = -2.45h$$

$$\text{EPE} = \frac{10 \times (0.0294 + h)^2}{2 \times 0.6} = \frac{25(0.0294 + h)^2}{3}$$

Energy is conserved. So, you have

$$\frac{25(0.0294 + h)^2}{3} - 2.45h = 0.0072$$

This gives $h^2 - 0.235h = 0$

and hence $h = 0$ or $h = 0.235$

The solution $h = 0$ corresponds to the starting position, A, so the pan falls by 0.235 m before coming instantaneously to rest at B.

$h^2 - 0.235\,h = 0.000864$ which is zero to 3 s.f.

M2

Example 11

One end of an elastic string, of natural length 1.2 m and modulus of elasticity 150 N, is fixed to a point A. A particle of mass 0.75 kg is attached to the other end. The particle is held at A and then released from rest. Find how far the particle falls before coming instantaneously to rest.

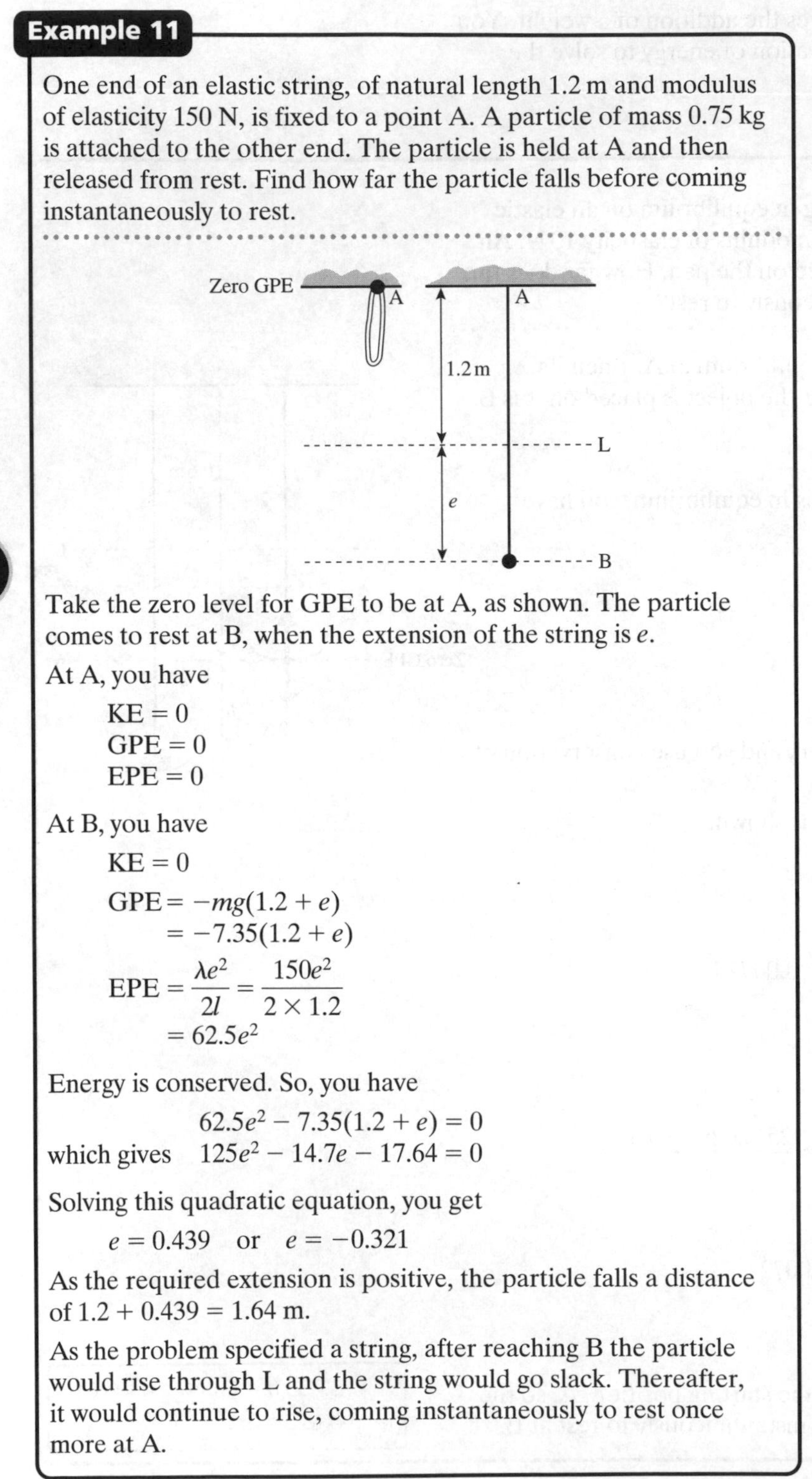

Take the zero level for GPE to be at A, as shown. The particle comes to rest at B, when the extension of the string is e.

At A, you have

$$\text{KE} = 0$$
$$\text{GPE} = 0$$
$$\text{EPE} = 0$$

At B, you have

$$\text{KE} = 0$$
$$\text{GPE} = -mg(1.2 + e)$$
$$= -7.35(1.2 + e)$$
$$\text{EPE} = \frac{\lambda e^2}{2l} = \frac{150e^2}{2 \times 1.2}$$
$$= 62.5e^2$$

Energy is conserved. So, you have

$$62.5e^2 - 7.35(1.2 + e) = 0$$

which gives $125e^2 - 14.7e - 17.64 = 0$

Solving this quadratic equation, you get

$$e = 0.439 \quad \text{or} \quad e = -0.321$$

As the required extension is positive, the particle falls a distance of $1.2 + 0.439 = 1.64$ m.

As the problem specified a string, after reaching B the particle would rise through L and the string would go slack. Thereafter, it would continue to rise, coming instantaneously to rest once more at A.

Note In Example 11, the second solution, $e = -0.321$ m, does not correspond to the starting point, A. To interpret it, you need to remember that the model does not distinguish between a string and a spring. Had you used a spring, the model predicts that, after coming instantaneously to rest at B, the particle would then rise above L, compressing the spring and coming to rest once more when the length of the spring reached $(1.2 - 0.321)$ m. It would be difficult to demonstrate this effect in practice, though if you were to start the motion by pulling the particle down to B it might be possible to achieve it.

M2

Example 12

Particles of mass 2 kg and 3 kg are attached to either end of a light, elastic string of natural length 2 m and modulus of elasticity 200 N. The particles are placed on a smooth horizontal surface. They are pulled to a distance of 5 m apart and released from rest. Find the speeds of the particles when the string goes slack and find the point at which the particles collide.

Let the final speeds of the particles be v_1 and v_2, as shown.

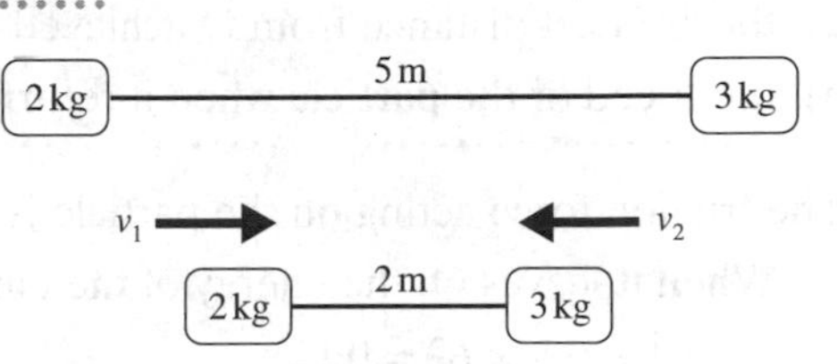

Initially, you have

$$\text{KE} = 0$$

$$\text{EPE} = \frac{\lambda e^2}{2l} = \frac{200 \times 3^2}{2 \times 2} = 450 \text{ J}$$

When string goes slack, you have

$$\text{KE} = \tfrac{1}{2} \times 2 \times v_1^2 + \tfrac{1}{2} \times 3 \times v_2^2 = \frac{2v_1^2 + 3v_2^2}{2}$$

$$\text{EPE} = 0$$

Energy is conserved. So, you have

$$\frac{2v_1^2 + 3v_2^2}{2} = 450$$

and so $2v_1^2 + 3v_2^2 = 900$ [1]

To solve the problem, you need a second equation in v_1 and v_2. To find this, you can use conservation of momentum, because all the forces involved in the problem are internal to the system.

Therefore, you have

Initially: momentum $= 0$
When string goes slack: momentum $= 2v_1 - 3v_2$
and hence $2v_1 - 3v_2 = 0$ [2]

From [2], you have $v_1 = \frac{3v_2}{2}$. Substituting this into [1], gives

$$2 \times \left(\frac{3v_2}{2}\right)^2 + 3v_2^2 = 900$$

This gives $v_2^2 = 120$ and hence $v_2 = 11.0$

Substituting back, you find $v_1 = 16.4$

So, the final speeds are 16.4 m s^{-1} for the 2 kg particle, and 11.0 m s^{-1} for the 3 kg particle.

As there are no external forces, the centre of mass of the system undergoes no acceleration. It was initially stationary at a point dividing the line joining the particles in the ratio 3 : 2. For it to remain stationary, the particles must collide at that point, which is at 3 m from the initial position of the 2 kg particle.

Example 13

A particle of mass 0.5 kg is attached to one end of an elastic string of natural length 1 m and modulus of elasticity 50 N. The other end of the string is attached to a fixed point O on a rough horizontal plane. The coefficient of friction between the particle and the plane is 0.4. The particle is projected from O along the plane with initial speed 6 ms^{-1}. Find

a) the greatest distance from O achieved by the particle

b) the speed of the particle when it returns to O.

The friction force acting on the particle is $F = 0.4 \times 0.5g = 1.96$ N.

a) When it leaves O, the energy of the particle is

$$\tfrac{1}{2} \times 0.5 \times 6^2 = 9 \text{ J}$$

When the string reaches it greatest extension, e, the EPE is

$$\frac{50e^2}{2 \times 1} = 25e^2$$

The work done against friction is $1.96(e + 1)$.

By the work–energy principle, you have

$$25e^2 + 1.96(e + 1) = 9$$

and hence $\quad 25e^2 + 1.96e - 7.04 = 0$

Solving this equation gives $e = 0.493$ or -0.571.
Clearly, the negative root is inappropriate, so the greatest distance from O achieved by the particle is 1.49 m.

b) When it returns to O, the particle has travelled $2 \times 1.493 = 2.986$ m against friction. Hence, the total work done against friction is $1.96 \times 2.986 = 5.852$ J.

The KE of the particle when it arrives back at O is, therefore,

$$9 - 5.852 = 3.148 \text{ J}.$$

If the speed of the particle is then v, you have

$$\tfrac{1}{2} \times 0.5v^2 = 3.148$$

and so $\quad v = 3.55$

So, the particle returns to O with a speed of 3.55 m s^{-1}.

M2

Exercise 5B

1 A ball of mass 500 g is fastened to one end of a light, elastic rope, whose unstretched length is 3 m and whose modulus of elasticity is 90 N. The other end of the rope is fastened to a bridge. The ball is held level with the fixed end, and is released from rest.

a) What will be the speed of the ball when it has fallen to the point where the rope is taut but unstretched?

b) How far below the bridge is the lowest point reached by the ball?

2 A ball of mass 0.4 kg is fastened to one end of a light, elastic string of natural length 4 m and modulus of elasticity 60 N. The other end is fastened to a point A on the rail of a high balcony. The ball is projected vertically downwards with initial speed 10 m s^{-1}. The ball comes instantaneously to rest at a point B vertically below A. It then rebounds, travelling past A and coming instantaneously to rest at a point C vertically above A. Find the lengths of AB and AC.

3 A particle of mass 2 kg is fastened to one end of a light, elastic string of natural length 1 m and modulus of elasticity 20 N. The other end of the string is fastened to a point A at the top of a smooth plane inclined at 30° to the horizontal. The particle is held at a point 3 m directly down the slope from A. The particle is released from rest. Find the speed at which the particle passes through A.

4 A catapult is made by fastening an elastic string of natural length 10 cm to points A and B, a distance of 6 cm apart. The modulus of elasticity of the string is 5 N. A stone of mass 10 g is placed at the centre of the string, which is then pulled back until the stone is 25 cm from the centre of AB. Find the greatest speed reached by the stone when it is released.

5 Find the work done in stretching a spring of natural length 2.5 m and modulus of elasticity 160 N from a length of 3 m to a length of 3.5 m.

6 A particle of mass 2 kg is suspended from a point A on the end of a spring of natural length 1 m and modulus of elasticity 196 N.

a) Find the length of the spring when the particle hangs in equilibrium.

The particle is now pulled down a distance of 0.5 m and released from rest.

b) Find the distance below A at which the particle next comes instantaneously to rest, assuming that the spring can compress to that point without the coils touching.

c) Find the highest position reached by the particle if, instead of the spring, you had used an elastic string of the same natural length and modulus.

7 A block, B, of mass 5 kg is fastened to one end of each of two springs. The other ends are fastened to two points, A and C, 4 m apart on a smooth, horizontal surface, as shown.

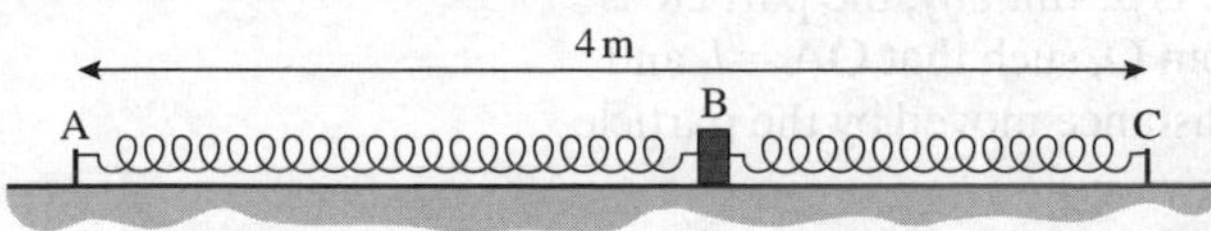

Spring AB has natural length 2 m and modulus of elasticity 30 N. Spring BC has natural length 1 m and modulus of elasticity 40 N.

a) Find the length AB when the block rests in its equilibrium position.
b) Find the tensions in the springs in this position.
c) Find the total elastic potential energy of the system in this position.

The block is moved 0.5 m towards C from the equilibrium position, and is held there.

d) What are the tensions or compressions in the springs in this position?
e) What is the total elastic potential energy of the system in this position?

The block is now released.

f) What is the speed of the block as it passes its equilibrium position?
g) How far beyond its equilibrium position does it travel before coming instantaneously to rest?

M2

8 An elastic string has natural length a and modulus of elasticity mg. Particles of masses m and $2m$ are attached to its ends. The particles are held at rest a distance $3a$ apart on a smooth, horizontal surface and are then released. Find the speeds of the particles at the moment when the string goes slack, and find the point where they collide.

9 A particle of mass 2 kg is attached to one end of an elastic string of natural length 1.2 m and modulus of elasticity 240 N. The other end of the string is fixed to a point A on a rough, horizontal plane. The particle is held at rest on the plane with the string stretched and is then released. The particle just reaches A before coming to rest. The coefficient of friction between the particle and the plane is 0.5.

a) Find the initial extension of the string.
b) Find the speed of the particle at the moment when the string goes slack.

10 A particle of mass 2 kg is attached to one end of an elastic string of natural length 2 m and modulus of elasticity 40 N. The other end of the string is attached to a fixed point O on a rough horizontal plane. The coefficient of friction between the particle and the plane is 0.5. The particle is projected from O along the plane with initial velocity v ms^{-1}. The particle returns and comes to rest exactly at O. Find

a) its furthest distance from O b) the value of v.

11 A particle of mass m is attached by means of a light elastic string of natural length l and modulus of elasticity λ to a fixed point O on a rough plane inclined at an angle α to the horizontal. The coefficient of friction between the plane and the particle is μ. Initially, the particle is held at point A directly down the slope from O, such that $\text{OA} = l$, and is then released from rest. Show that the distance moved by the particle before coming to rest again is

$$\frac{2mgl(\sin\alpha - \mu\cos\alpha)}{\lambda}$$

Summary

You should know how to ...	Check out
1 Use Hooke's law to find the tension in elastic strings and springs.	**1** An elastic string has modulus $5mg$ N and natural length $3a$. Find the tension in the string when it is stretched to a total length of $5a$.
2 Calculate elastic potential energy.	**2** Find the elastic potential energy of the string in question 1.

Revision exercise 5

1 A block, of mass 4 kg, is attached to one end of a length of elastic string. The other end of the string is fixed to a wall. The block is placed on a horizontal surface as shown in the diagram below.

The elastic string has natural length 60 cm and modulus of elasticity 60 N. The block is pulled so that it is 1 m from the wall and is then released from rest.

a) Calculate the elastic potential energy when the block is 1 m from the wall.

b) If the surface is smooth, show that the speed of the block when it hits the wall is 2 m s^{-1}.

c) The surface is in fact rough and the coefficient of friction between the block and the surface is 0.3.

i) Show that the speed of the block when the string becomes slack is approximately 1.28 m s^{-1}.

ii) Determine whether or not the block will hit the wall. *(AQA, 2002)*

2 An elastic rope has natural length 4 m and modulus of elasticity 80 N. A particle, of mass 2 kg, is attached to one end of the rope, and the other end is fixed at the point A. The particle is released from rest at A and falls vertically.

a) When the rope just becomes taut, find:

i) the kinetic energy of the particle;

ii) the speed of the particle.

b) i) The maximum extension of the rope during the motion is x metres. Show that x satisfies the equation

$$10x^2 - 19.6x - 78.4 = 0$$

ii) Hence find the maximum length of the rope.

c) State clearly **one** important assumption that you have made. *(AQA, 2004)*

3 A bungee jumper, of mass 80 kg, is attached to an elastic rope of natural length 20 m and modulus of elasticity 2000 newtons. The other end of the elastic rope is attached to a bridge. The bungee jumper steps off the bridge at the point where the rope is attached and falls vertically. When the bungee jumper has fallen x m, his speed is v m s^{-1}.

a) By considering energy, show that when x exceeds a certain minimum value

$$20v^2 = 1392x - 25x^2 - 10\,000$$

and state this minimum value of x.

b) Find the maximum value of x.

c) i) Show that the speed of the bungee jumper is a maximum when $x = 27.84$ m.

ii) Hence find the maximum speed of the bungee jumper. *(AQA, 2002)*

M2

4 An elastic string has natural length 2 m and modulus of elasticity λ newtons. One end of the string is fixed at the point O, and a particle of mass 20 kg is attached to the other end of the string.

a) When in equilibrium the particle is 2.7 m below O. Show that $\lambda = 560$.

b) The particle is now held at O and released from rest. The maximum length of the string in the subsequent motion is L.

i) Show that L satisfies the equation

$$5L^2 - 27L + 20 = 0$$

ii) Find the maximum length of the string. *(AQA, 2004)*

5 A bungee jumper, of mass 70 kg, is attached to one end of a light elastic cord of natural length 14 m and stiffness 196 N m^{-1}. The other end of the cord is attached to a bridge, approximately 40 metres above a river.

The bungee jumper steps off the bridge at the point where the cord is attached and falls vertically. The bungee jumper can be modelled as a particle throughout the motion. Hooke's law can be assumed to apply throughout the motion.

a) Find the speed of the bungee jumper at the instant the cord first becomes taut.

b) The cord extends by x metres beyond its natural length before the bungee jumper first comes instantaneously to rest.

i) Show that $x^2 - 7x - 98 = 0$.

ii) Hence find the value of x.

iii) Calculate the deceleration experienced by the bungee jumper at this point. *(AQA, 2003)*

6 Circular motion

This chapter will show you how to

- ✦ Convert between different units for angular speed
- ✦ Know and use the relationships $v = r\omega$, $a = r\omega^2$, $a = \frac{v^2}{r}$
- ✦ Find the position, velocity and acceleration vectors of a particle moving in a circle
- ✦ Model the motion of a particle moving in a circle with constant speed
- ✦ Model the motion of a particle moving as a conical pendulum
- ✦ Model the motion of a particle moving in a vertical circle

Before you start

M2

You should know how to ...	Check in
1 Use Hooke's law to find the tension in elastic strings and springs.	**1** An elastic string has modulus $6mg$ N and natural length $4a$ m. Find the tension in the string when it is stretched by $2a$ m.
2 Use the principle of conservation of energy.	**2** A body of mass 6 kg travelling with speed 4 m s^{-1} starts to slide down a slope. Find its speed after the body has dropped a vertical distance of 5 m.
3 Use and convert between radians and degrees.	**3** a) Convert 45° into radians. b) Convert $\frac{\pi}{6}$ radians into degrees.
4 Differentiate terms involving vectors.	**4** $\mathbf{r} = 3\cos 4t\mathbf{i} + 3\sin 4t\mathbf{j} - 6t\mathbf{k}$ Find a) $\dot{\mathbf{r}}$ b) $\ddot{\mathbf{r}}$
5 Use $\mathbf{F} = m\mathbf{a}$	**5** A force of $(8\mathbf{i} + 8\mathbf{j} - 6\mathbf{k})$ N acts on a particle of mass 2 kg. Find the acceleration of the body.
6 Resolve forces.	**6** The vector $\mathbf{F}$ has magnitude 6 N and acts in a direction 25° to the $\mathbf{i}$ direction. Find the vector in the form $\lambda\mathbf{i} + \mu\mathbf{j}$.

Circular motion is a significant aspect of motion in two dimensions. Many objects move in a circular path: for example, a car on a roundabout, a pendulum on a clock or a sock in a spin dryer.

6.1 Linear and angular speed

The motion of a particle travelling in a circle can be described in terms of its displacement, speed and acceleration along the arc of the circle. However, it is often more appropriate to describe it in terms of the angle subtended at the centre of the circle by its path.

For example, consider points A and B on an old vinyl record revolving on a turntable at 33 revolutions per minute (the preferred abbreviation for which is rev min^{-1}, although rpm is more widely used). Suppose that A is 5 cm and B is 10 cm from the centre of the disc.

In 1 minute:	A travels $33 \times 2\pi \times 0.05 = 10.4$ m
	B travels $33 \times 2\pi \times 0.10 = 20.7$ m
So you have	speed of A $= 10.4 \div 60 = 0.173\ \text{m s}^{-1}$
and	speed of B $= 20.7 \div 60 = 0.346\ \text{m s}^{-1}$

M2

This means that points on the disc have different linear speeds depending on their distance from the centre. To describe the rate at which the disc is rotating, it is better to consider its angle of rotation.

In 1 rotation the disc turns through an angle of 2π radians. Hence, in 1 minute it turns through $33 \times 2\pi = 66\pi$ radians. Therefore, the angular speed of the disc is

$$66\pi \div 60 = 1.1\pi \text{ radians per second (written } 1.1\pi \text{ rad s}^{-1})$$

6.2 Angular displacement, speed and acceleration

The **angular displacement**, θ, of a particle P is the angle in radians rotated by the radius OP from some reference position (usually the position when time is zero). You could, in theory, use degrees, but, because you will need to use calculus methods, radians should **always** be used.

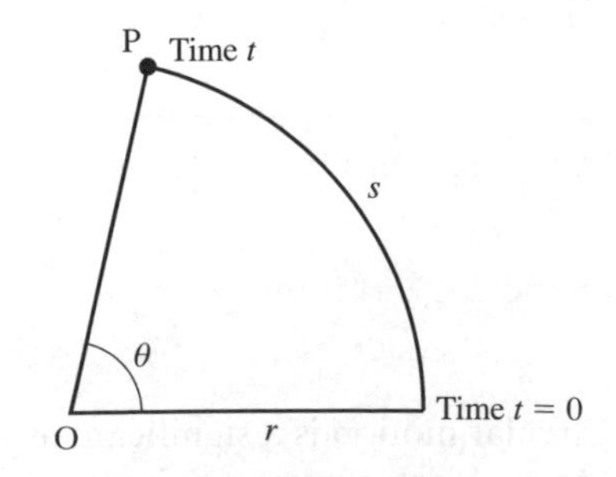

The rate of change of the angular displacement gives the **angular speed** (in rad s^{-1}). This is usually denoted by ω.

Thus you have

$$\frac{d\theta}{dt} = \dot{\theta} = \omega$$

The rate of change of the angular speed gives the **angular acceleration** (in rad s^{-2}).

Thus you have

$$\frac{d^2\theta}{dt^2} = \ddot{\theta} = \dot{\omega}$$

6.3 Relation between linear and angular measures

If a particle travels along the arc shown in the diagram above right, it defines a sector with angle θ, radius r and arc length s. We know that

θ must be in radians.

$$s = r\theta$$

As r is constant, differentiating this formula gives

$$\frac{ds}{dt} = r\frac{d\theta}{dt}$$

If the speed of the particle is v, this becomes

$$v = r\dot{\theta} \quad \text{or} \quad v = r\omega$$

Differentiating again gives

$$\frac{dv}{dt} = r\ddot{\theta} = r\dot{\omega}$$

M2

Example 1

A particle travels round the circumference of a circle of radius 6 m at a rate of 30 rev min^{-1}. Find a) its angular speed in rad s^{-1} and b) its linear speed around the circle.

a) One revolution $= 2\pi$ rad
The particle has angular speed $2\pi \times 30 = 60\pi$ rad min^{-1}
$= 60\pi \div 60 = \pi$ rad s^{-1}

b) Using $v = r\omega$, you have
$v = 6 \times \pi = 6\pi \text{ m s}^{-1}$ or 18.8 m s^{-1}

Exercise 6A

1 Convert an angular speed of
a) 15 rev min^{-1} to rad s^{-1} b) 25 rad s^{-1} to rev min^{-1}.

2 Find the speed in m s^{-1} of a particle moving in a circle of radius
a) 60 cm at 15 rad s^{-1} b) 2.5 m at 5π rev min^{-1}.

3 Mobusar and Samantha are on a roundabout. Their distances from the centre of the roundabout are 1.3 m and 1.8 m respectively. Mobusar is travelling at 3.5 m s^{-1}. Find
a) the angular speed of the roundabout in rad s^{-1}
b) Samantha's speed in m s^{-1}.

4 Find the angular speed in rad s^{-1} of a bicycle wheel of radius 35 cm when the bicycle is travelling at 18 km h^{-1}.

7 The hands of a large public clock are 70 cm and 110 cm long. Find in m s^{-1} the speed of the tip of each hand.

6.4 Modelling circular motion

As with other forms of motion, you will need to make various modelling assumptions for circular motion in different situations. However, the following considerations will underlie all your models:

- Velocity is a vector quantity, which is changed if either its magnitude or its direction is altered.
- When a body moves along a circular path, its velocity at any instant is directed along the tangent to the circle at that point. This direction is constantly changing, which means that its velocity is changing.
- A change in velocity means that the body is accelerating.
- By Newton's second law, there must be a force acting on the body to cause this acceleration.

Motion with uniform speed

The simplest modelling assumptions are:

- The object is a particle.
- The object is travelling around a circle at uniform speed.

Consider the consequences of these two assumptions.

A particle, is moving in a circle, radius r and centre O, with constant angular speed ω rad s^{-1}.

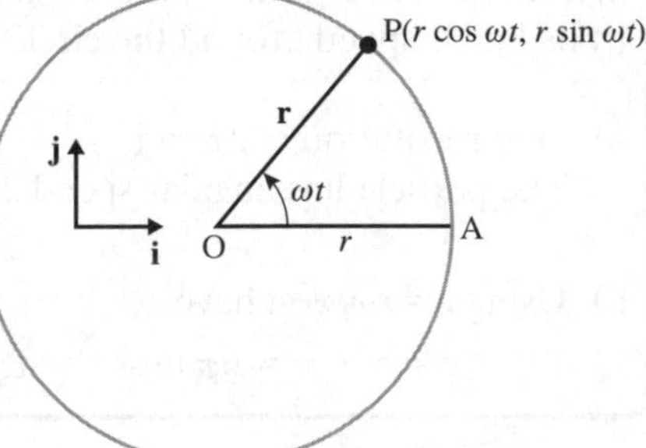

The particle is initially at point A. After a time t s, the particle is at P. Its angular displacement, the angle AOP in the diagram, is then ωt.

Taking O as the origin and with unit vectors as shown, the coordinates of P are $(r\cos\omega t, r\sin\omega t)$ and its position vector is therefore given by

$$\mathbf{r} = r\cos\omega t\mathbf{i} + r\sin\omega t\mathbf{j} \qquad [1]$$

You find the velocity of the particle by differentiating $\mathbf{r}$ with respect to t. This gives

$$\mathbf{v} = -r\omega\sin\omega t\mathbf{i} + r\omega\cos\omega t\mathbf{j} \qquad [2]$$

Similarly, you find the acceleration of the particle by a second differentiation. This gives

$$\mathbf{a} = -r\omega^2\cos\omega t\mathbf{i} - r\omega^2\sin\omega t\mathbf{j} \qquad [3]$$

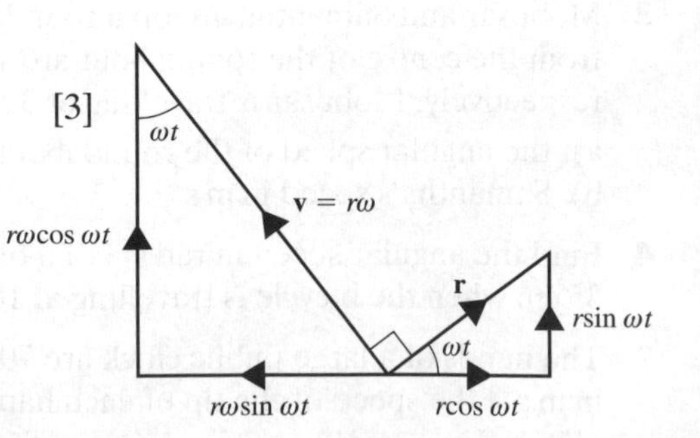

The diagram on the right illustrates the vectors $\mathbf{r}$ and $\mathbf{v}$. This provides confirmation that

- the magnitude of the velocity is $r\omega$
- the velocity vector is perpendicular to the position vector. In other words, the velocity of the body has a direction **tangential to its circular path**.

From equations [1] and [3], you can see that

$$\mathbf{a} = -\omega^2\mathbf{r}$$

The acceleration therefore acts in the opposite direction to **r**. Since **r** is directed away from the centre of the circle, it follows that the acceleration acts towards the centre of the circle, whatever the position of P on the circle.
You also have $|\mathbf{a}| = |-\omega^2\mathbf{r}|$
That is

$$a = r\omega^2 \qquad [4]$$

Since ω and r are constants, the magnitude of the acceleration is constant, but its direction is constantly changing.

An alternative form of the expression for the magnitude of the acceleration can be obtained as follows:

You know that $v = r\omega$ and so $\omega = \dfrac{v}{r}$

Substituting this in equation [4], you get

$$a = \frac{v^2}{r}$$

Summary
For a particle moving in a circle of radius r with uniform angular speed ω:

- The velocity is tangential to the circle.
- The speed around the circle is $v = r\omega$.
- The acceleration acts towards the centre of the circle.
- The magnitude of the acceleration is $r\omega^2$ or $\dfrac{v^2}{r}$

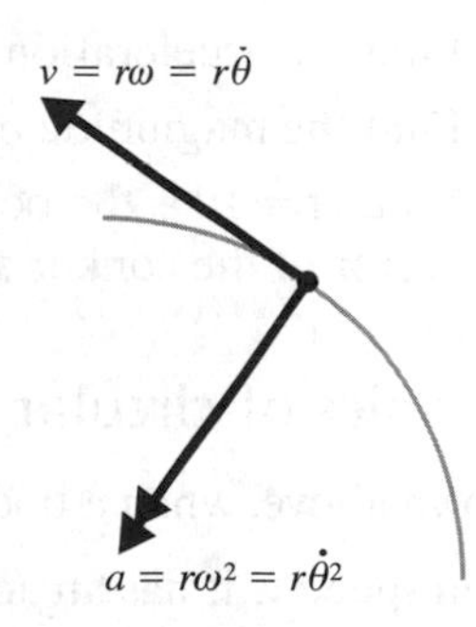

M2

Exercise 6B

1 Find the magnitude of the acceleration of a particle travelling
a) around a circle of radius 2.5 m at a uniform speed of 8 m s^{-1}
b) around a circle of radius 6 m at a uniform speed of 20 m s^{-1}
c) around a circle of diameter 16 m at a uniform speed of 7.2 km h^{-1}
d) around a circle of radius 4.2 m with uniform angular speed 3 rad s^{-1}
e) around a circle of diameter 20 m with uniform angular speed 0.5 rad s^{-1}
f) around a circle of diameter 2.6 m at a constant 12 rev min^{-1}.

2 The Earth rotates once in every 24 hours (approximately). Its radius is approximately 6400 km.

a) Find the angular speed of the Earth in rad s^{-1}.

b) Find the acceleration of a person standing on the equator.

c) At what latitude would the person's acceleration be 0.02 m s^{-2}?

d) What would be the direction of the acceleration of the person in part c)?

3 The Earth orbits the Sun once every 365 days (approximately). Assuming that its orbit is circular, with a radius of 1.49×10^6 km, find its acceleration.

4 A car is travelling round a roundabout of radius 10 m with a uniform speed of 6 m s^{-1}.

a) Find its angular speed, ω.

b) Taking the centre of the roundabout as the origin, the car is initially at the point with position vector $10\mathbf{i}$ and is moving in an anticlockwise direction. Find an expression for the position vector, $\mathbf{r}$, of the car at time t.

c) Differentiate the position vector to find expressions for the velocity and acceleration vectors of the car.

M2

5 A boy is swinging a conker so that it moves in a horizontal circle above his head with uniform speed. At time t, the position vector of the conker is given by

$$\mathbf{r} = (0.3 \cos 10t\mathbf{i} + 0.3 \sin 10t\mathbf{j}) \text{ m}$$

a) Find the acceleration vector of the conker.

b) Find the magnitude of the conker's acceleration.

c) When $t = 10$ s, the boy releases the string. What is the velocity vector of the conker at this time?

Mechanics of circular motion

As shown above, when a body is travelling in a circle of radius r with uniform speed v, it has an acceleration of magnitude $\frac{v^2}{r}$, or $r\omega^2$, towards the centre of the circle, where ω is the uniform angular speed.

As a consequence of Newton's second law, there must be a force acting to cause the acceleration. A body on which no force acts will move in a straight line. Indeed, if a body travelling in a circle suddenly has the accelerating force removed (such as when a conker string breaks), the body will move along the tangent at that point.

In solving a practical problem, you need to identify all the forces acting and then write down the equation of motion. Because the acceleration acts towards the centre of the circle, it must be in the plane of the circle. There can be no acceleration perpendicular to this plane.

M2

Example 2

A particle of mass 3 kg is attached to a light, inextensible string of length 2 m, which is fastened to a point on a smooth table. The particle is set in motion in such a way that it describes circles around the fixed point with a speed of 6 m s^{-1}. Find the tension in the string.

For the motion described to be possible, the string must be taut. The radius of the circle is therefore 2 m.

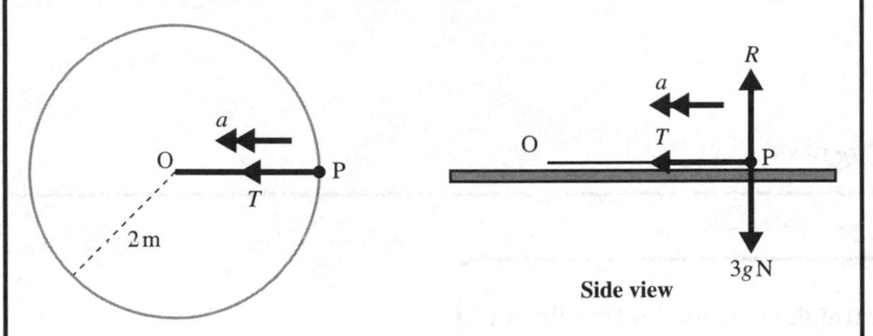

There is no acceleration vertically and the tension has no vertical component. Hence you only need consider the equation of motion for the horizontal direction.

The acceleration towards the centre of the circle is

$$\frac{v^2}{r} = \frac{6^2}{2} = 18 \text{ m s}^{-2}$$

Resolving in the direction PO and using Newton's second law ($F = ma$), you have the equation of motion

$$T = 3 \times 18 = 54$$

So, the tension in the string is 54 N.

Example 3

Two particles, each of mass m kg, are attached to the ends of a light, inextensible string which passes through a hole in a smooth table. One particle moves on the table in a circular path of radius r m around the hole, so that the string is taut. The other particle hangs freely and at rest. What is the speed of the particle on the table?

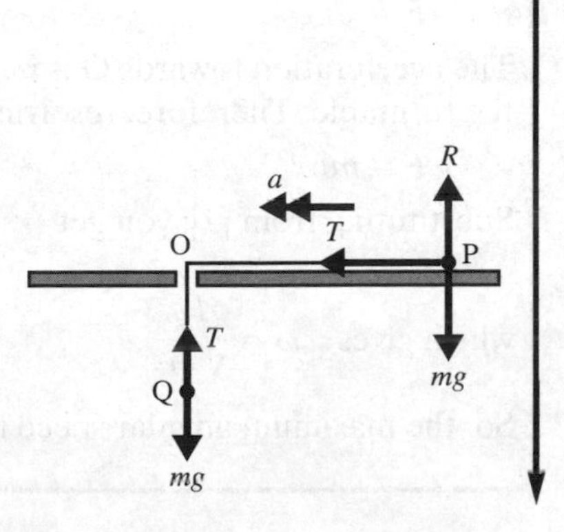

Particle P is moving in a circular path and Q is stationary. The tension is the same throughout the string.

Resolving vertically for Q, you have

$$T - mg = 0$$

and hence $T = mg$ [1]

P has no acceleration vertically, so you only need consider its equation of motion horizontally.

The acceleration of P is $\frac{v^2}{r}$ towards O.

Resolving for P along PO, you have

$$T = \frac{mv^2}{r} \quad [2]$$

Substituting for T from [1] into [2], you have

$$mg = \frac{mv^2}{r}$$

which gives $v^2 = rg$

and hence $v = \sqrt{rg}$

So, the velocity of the particle P is $\sqrt{rg}$ m s^{-1}.

Example 3 is a 'classic' problem and the situation could not be achieved in practice. Friction between the particle and the table, and between the string and the edge of the hole, would quickly change the speed.

M2

Example 4

A particle of mass m is placed on a rough horizontal turntable at a distance of 0.4 m from its centre, O. The coefficient of friction between the particle and the turntable is $\frac{1}{4}$. What is the maximum angular speed at which the turntable can move if the particle is not to slide?

a

R

O

F

P

0.4 m

mg N

Resolving vertically, you obtain

$$R - mg = 0$$

and so $R = mg$ [1]

If the particle is on the point of sliding, you have

$$F = \tfrac{1}{4}R$$

So, substituting from [1], you get

$$F = \tfrac{1}{4}mg \quad [2]$$

The acceleration towards O is $r\omega^2$, where ω is the angular speed of the turntable. Therefore, resolving in the direction PO, you have

$$F = mr\omega^2 \quad [3]$$

Substituting from [2], you get

$$\tfrac{1}{4}mg = mr\omega^2$$

which gives $\omega = \sqrt{\frac{g}{4r}}$

So, the maximum angular speed is $\sqrt{\frac{g}{4r}}$.

Satellites in circular orbits

According to Newton's law of Universal Gravitation, the gravitational force between objects of masses m_1 and m_2 separated by a distance d is given by

> You are not required to know this formula. It is introduced here to give another example of motion in a circle.

$$F = \frac{Gm_1m_2}{d^2}$$

where $G = 6.67 \times 10^{-11}\ \text{N m}^2\ \text{kg}^{-2}$ is the universal gravitational constant.

The mass of the Earth is approximately 5.98×10^{24} kg. Hence, a satellite of mass m kg at a distance r m from the centre of the Earth is attracted towards the Earth by a force

$$F = \frac{6.67 \times 10^{-11} \times 5.98 \times 10^{24} m}{r^2} = \frac{3.99 \times 10^{14} m}{r^2}\ \text{N}$$

In general, the orbit of a satellite is elliptical, but many satellites are placed in orbits which are approximately circular. In particular, it is useful for communications and related purposes to have a satellite moving in a circular orbit so that it remains stationary relative to the surface of the Earth. This is called a **parking** or **geostationary** orbit.

M2

Suppose that a satellite of mass m kg moves with an angular velocity of ω rad s^{-1} in a circular orbit of radius r m (where r is obviously greater than the radius of the Earth). From Newton's second law, you have

$$\frac{3.99 \times 10^{14} m}{r^2} = mr\omega^2$$

It is then possible to find the value of ω and hence the period of the orbit, for a given value of r.

Example 5

Given that the radius of the Earth is 6.37×10^6 m, find the orbital period of a satellite in a circular orbit at a height of 500 km above the surface of the Earth.

Let the mass of the satellite be m and its angular velocity be ω.

The radius of the orbit is $6.37 \times 10^6 + 500\,000 = 6.87 \times 10^6$ m.

The gravitational force acting on the satellite is

$$\frac{6.67 \times 10^{-11} \times 5.98 \times 10^{24} m}{(6.87 \times 10^6)^2} = 8.45m\ \text{N}$$

By Newton's second law, you have

$$8.45m = 6.87 \times 10^6 m\omega^2$$

and so

$$\omega = 0.0011\ \text{rad s}^{-1}$$

The period of the orbit is, therefore, $\frac{2\pi}{0.0011} = 5665$ s or 1 h 34 min to the nearest minute.

Exercise 6C

1 A particle of mass 2 kg is attached by a light, inextensible string of length 1.2 m to a fixed point on the surface of a horizontal, smooth table. The particle travels in a circle on the table at a speed of 2.5 m s^{-1}. Find the tension in the string.

2 A railway engine of mass 80 tonnes is travelling in an arc of a horizontal circle of radius 150 m at a speed of 72 km h^{-1}. What total sideways force is being exerted on the wheels by the rails?

3 A string of length 80 cm can just support a suspended mass of 40 kg without breaking. A 2 kg mass is attached to the string and the other end of the string is fastened to a point on the surface of a smooth, horizontal table. The mass is made to move in a circle on the table. Find, in rev min^{-1}, the maximum rate at which the mass can revolve without breaking the string.

4 A coin of mass 0.005 kg is placed on a turntable, which is then rotated at 30 rev min^{-1}. If the coin does not slip, what is the minimum coefficient of friction when

a) the coin is placed at a distance of 6 cm from the centre

b) the coin is placed at a distance of 8 cm from the centre

c) a coin of mass 0.01 kg is used in each of the cases a) and b)?

5 A railway engine of mass 50 tonnes is rounding a horizontal curve of radius 200 m. The track can withstand a sideways force of 36 kN before it buckles.

a) What is the maximum speed at which the train can travel safely?

b) What would be the maximum speed for an engine of twice the mass?

6 Four particles, each of mass 3 kg, are connected by light, inextensible strings, each 0.08 m long, so that they form a square with the particles at the corners and the strings forming the sides. The particles are placed in this configuration symmetrically on a smooth turntable, which is made to rotate with angular speed 2 rad s^{-1}. Find the tension in the strings.

7 A rough, horizontal disc rotates at 2 rev s^{-1}. A particle is to be placed on the disc. The coefficient of friction between the particle and the disc is μ. Find, in terms of μ, the maximum distance from the centre at which the particle can be placed without its slipping.

8 A particle of mass 3 kg is placed on a rough, horizontal turntable and is connected to its centre by a light, inextensible string of length 0.8 m. The coefficient of friction between the particle and the turntable is 0.4. The turntable is made to rotate at a uniform speed. The tension in the string is 50 N. Find the angular speed of the turntable.

9 A particle, A, of mass m, is travelling in a horizontal circle on the surface of a smooth, horizontal table, on the end of a light, inextensible string, OA, of length $3a$, which is fastened to a point O on the table. The angular speed of the particle is ω. The string then catches on a peg, P, which is at a distance of $2a$ from O. Find the new angular speed of the particle and the new tension in the string.

10 Two particles, each of mass m, are attached to the ends of a light, inextensible string. The string passes through a hole in the centre of a rough, horizontal turntable. One particle is placed on the turntable at a distance a from its centre, and the other hangs freely below the turntable. The coefficient of friction between the particle and the turntable is μ. The contact between the string and the hole is smooth. The turntable is rotating with angular speed ω. Find the maximum and minimum values of ω if the particle on the turntable does not slip.

11 Newton's law of gravitation states that the force of attraction between two bodies of masses M and m is given by

$$F = \frac{GMm}{r^2}$$

where $G = 6.67 \times 10^{-11}\,\text{N m}^2\,\text{kg}^{-2}$ is the universal gravitational constant and r is the distance between the centres of the bodies.

a) A communications satellite is in geostationary orbit above the equator (that is, it moves in a circular orbit so as to always be above the same point on the Earth). What is the angular speed of the satellite?

b) Taking the mass of the Earth to be 5.97×10^{24} kg, find the radius of the satellite's orbit.

c) A second satellite moves in a circular orbit at a height of 1200 km above the surface of the Earth. Find the time taken (to the nearest minute) for it to complete one orbit.

12 A fairground ride consists of a large hollow cylinder placed with its axis vertical. The customers stand against the inside wall of the cylinder, and the whole assembly rotates. When it reaches top speed, the floor is lowered, leaving the riders supported only by the friction between themselves and the wall. The radius of the cylinder is 3 m and the coefficient of friction between a rider and the wall is $\frac{1}{3}$. What is the minimum angular speed at which the ride will operate effectively?

6.5 Problems involving non-horizontal forces

In all the situations so far examined, the forces causing the acceleration towards the centre of the circle have been horizontal. There are many situations where a body travels in a horizontal circle but where the constraining force has a vertical component.

Conical pendulum

The term 'conical pendulum' refers to the situation in which a body is attached by a string to a fixed point, and travels in a horizontal circle below that point. It is usual to make the modelling assumptions that the string is light and inextensible, that the body is a particle and that air resistance can be neglected.

M2

Example 6

A pendulum bob, P, of mass 4 kg, hangs at the end of a light, inextensible string of length 5 m. The other end of the string is fixed to a point O. The bob is made to describe a horizontal circle of radius 3 m with uniform speed. Find the tension in the string and the angular speed of the bob.

By Pythagoras's theorem it is clear that the circle is 4 m below point O. Hence, if θ is the angle between the string and the horizontal, you have

$$\cos\theta = \tfrac{3}{5} \text{ and } \sin\theta = \tfrac{4}{5}$$

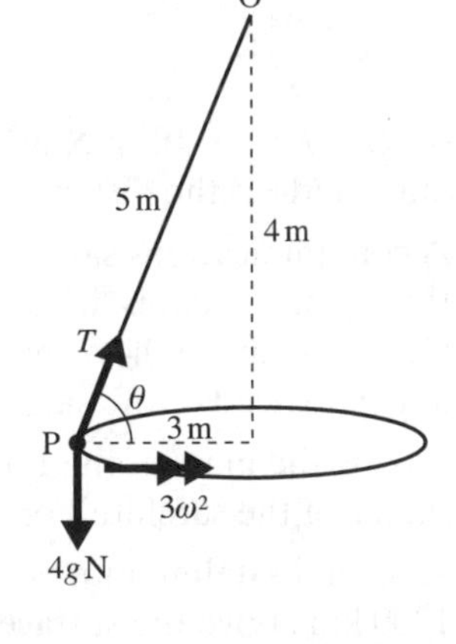

Let the tension in the string be T N and the angular speed be ω rad s^{-1}.

The bob has zero vertical acceleration, and an acceleration of $r\omega^2 = 3\omega^2$ towards the centre of the circle.

Resolving vertically, you have

$$T\sin\theta - 4g = 0$$

which gives $\tfrac{4}{5}T = 4g$

and hence $T = 5g = 49$ [1]

So the tension in the string is 49 N.

Resolving towards the centre of the circle, you have

$$T\cos\theta = 4 \times 3\omega^2$$

which gives $\omega^2 = \tfrac{1}{20}T$

Substituting from [1], you have

$$\omega^2 = 2.45$$

and so $\omega = 1.57$

So, the angular speed of the bob is 1.57 rad s^{-1}.

M2

Example 7

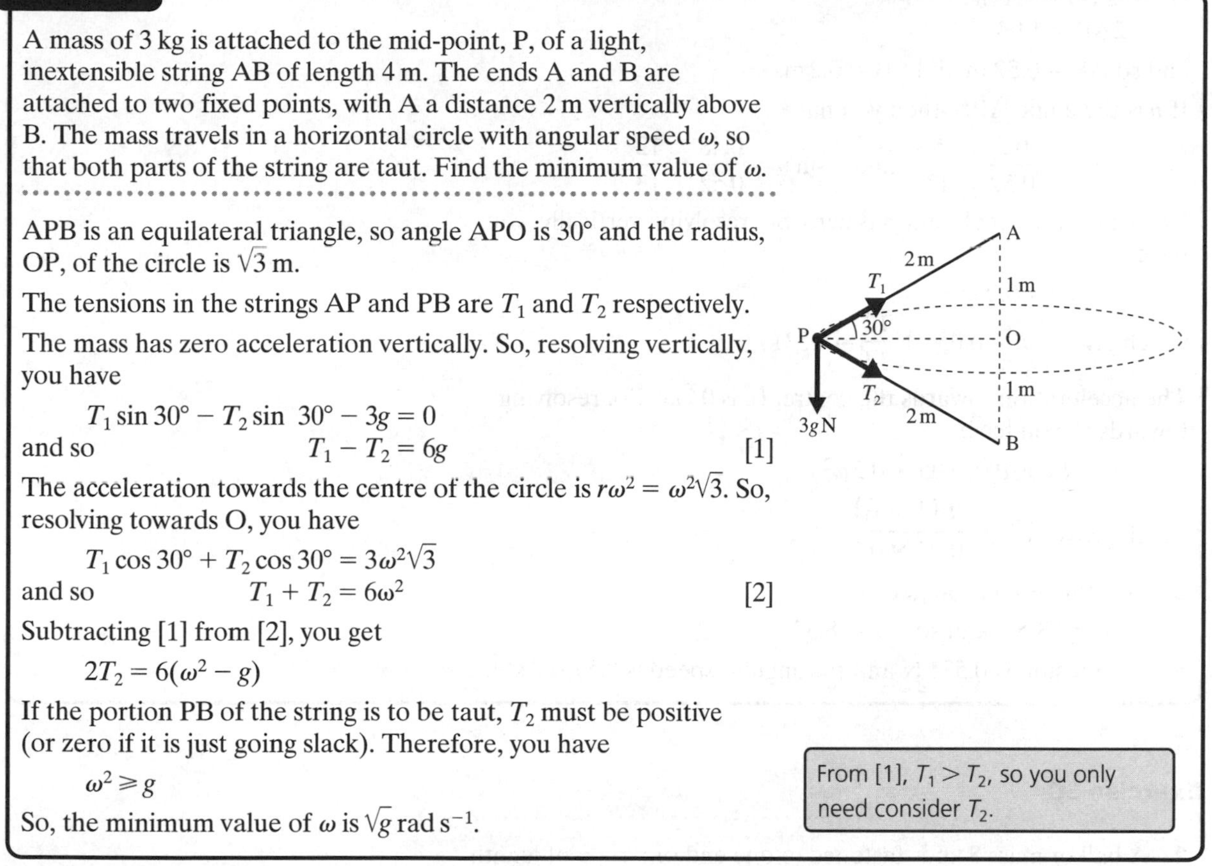

A mass of 3 kg is attached to the mid-point, P, of a light, inextensible string AB of length 4 m. The ends A and B are attached to two fixed points, with A a distance 2 m vertically above B. The mass travels in a horizontal circle with angular speed ω, so that both parts of the string are taut. Find the minimum value of ω.

APB is an equilateral triangle, so angle APO is 30° and the radius, OP, of the circle is $\sqrt{3}$ m.

The tensions in the strings AP and PB are T_1 and T_2 respectively.

The mass has zero acceleration vertically. So, resolving vertically, you have

$$T_1 \sin 30° - T_2 \sin 30° - 3g = 0$$

and so $\quad T_1 - T_2 = 6g \qquad [1]$

The acceleration towards the centre of the circle is $r\omega^2 = \omega^2\sqrt{3}$. So, resolving towards O, you have

$$T_1 \cos 30° + T_2 \cos 30° = 3\omega^2\sqrt{3}$$

and so $\quad T_1 + T_2 = 6\omega^2 \qquad [2]$

Subtracting [1] from [2], you get

$$2T_2 = 6(\omega^2 - g)$$

If the portion PB of the string is to be taut, T_2 must be positive (or zero if it is just going slack). Therefore, you have

$$\omega^2 \geqslant g$$

From [1], $T_1 > T_2$, so you only need consider T_2.

So, the minimum value of ω is $\sqrt{g}$ rad s^{-1}.

Example 8

A light, inextensible string, of length 0.72 m, is attached to two points A and B, where A is vertically above B and AB = 0.48 m. A smooth ring, P, of mass 0.05 kg, is threaded on the string and is made to move in a horizontal circle about B. Find the angular speed of the ring and the tension in the string.

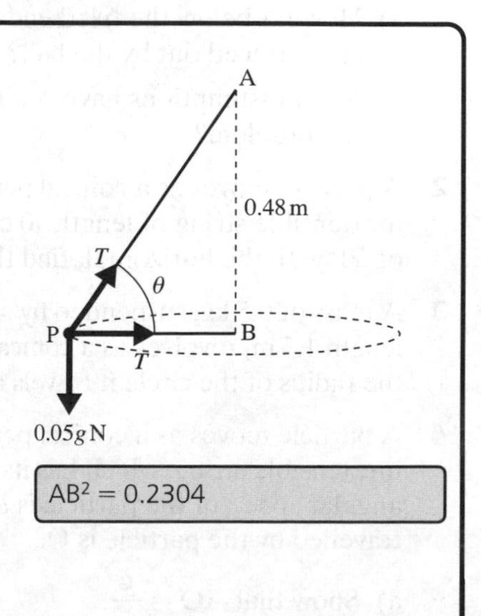

The first task is to find the lengths AP and BP.

Because the ring can move freely on the string, the tension in both portions of the string is the same.

You are given that $\quad$ AP + PB = 0.72 $\qquad [1]$

By Pythagoras's theorem, you have

$$AP^2 - PB^2 = 0.48^2 = 0.2304$$

$AB^2 = 0.2304$

and so $\quad (AP + PB)(AP - PB) = 0.2304$

Substituting from [1], you get

$$0.72(AP - PB) = 0.2304$$

and so $\quad AP - PB = 0.32 \qquad [2]$

Adding [1] and [2], you get

$$2AP = 1.04$$

and so AP = 0.52 m and PB = 0.2 m.

If θ is the angle APB, then you have

$$\cos\theta = \frac{0.2}{0.52} = \frac{5}{13} \quad \text{and} \quad \sin\theta = \frac{0.48}{0.52} = \frac{12}{13}$$

Vertically, the acceleration is zero. So, resolving vertically, you have

$$T\sin\theta = 0.05g$$

which gives $T = 0.05g \div \frac{12}{13} = 0.531$

The acceleration towards the centre, B, is $0.2\omega^2$. So, resolving towards B, you have

$$T + T\cos\theta = 0.05 \times 0.2\omega^2$$

which gives $\omega^2 = \dfrac{T\left(1 + \frac{5}{13}\right)}{0.05 \times 0.2}$

Substituting for T, you get

$$\omega^2 = 73.5 \quad \text{and so} \quad \omega = 8.57$$

So, the tension is 0.531 N and the angular speed is 8.57 rad s^{-1}.

M2

Exercise 6D

1 A ball of mass 3 kg is fastened to one end of a rope of length 0.5 m. The other end of the rope is fixed and the ball rotates as a conical pendulum at a rate of 5 rad s^{-1}.

a) How far below the fixed end of the rope is the centre of the circle traced out by the ball?

b) What assumptions have you made in forming your solution to the problem?

2 A particle moves as a conical pendulum at the end of a light, inextensible string of length 40 cm. If the string makes an angle of 30° with the horizontal, find the angular speed of the particle.

3 A mass of 0.5 kg, suspended by a light, inextensible string of length 1.5 m, revolves as a conical pendulum at 30 rev min^{-1}. Find the radius of the circle it travels and the tension in the string.

4 A particle moves as a conical pendulum at the end of a light, inextensible string, which has its fixed end at point A. The angular speed of the particle is ω. The centre of the circle travelled by the particle is O.

a) Show that $AO = \dfrac{g}{\omega^2}$.

b) Explain why the string cannot be horizontal.

5 A ball of mass 1 kg is fastened to one end of a light, inextensible string of length 1 m. The ball is placed on a smooth, horizontal table. The string is suspended from a point above the table so that the particle moves as a conical pendulum whilst in contact with the table. The radius of the circle travelled by the particle is 0.5 m.

a) If the speed of the particle is 1.5 m s^{-1}, what is the normal reaction between the ball and the table?

b) What is the maximum speed the ball could travel without lifting off the table?

6 Two points, A and B, are on a vertical pole, 9 m apart with A above B. A rope of length 27 m is fastened at its ends to A and B. A smooth, heavy metal ring of mass m is threaded onto the rope. The ring is made to move in a horizontal circle about the pole. The upper section, AS, of the rope makes an angle θ with the vertical, as shown. Find the speed of the ring and the tension in the rope when

a) $\tan \theta = \frac{8}{15}$

b) $\tan \theta = \frac{4}{3}$

7 A ball, B, of mass 2 kg is attached to one end of a light, inextensible string. The string passes through a smooth, fixed ring, O, and a second ball, A, of mass 4 kg, is attached to the other end. B is made to move as a conical pendulum while A hangs vertically below the ring, as shown.

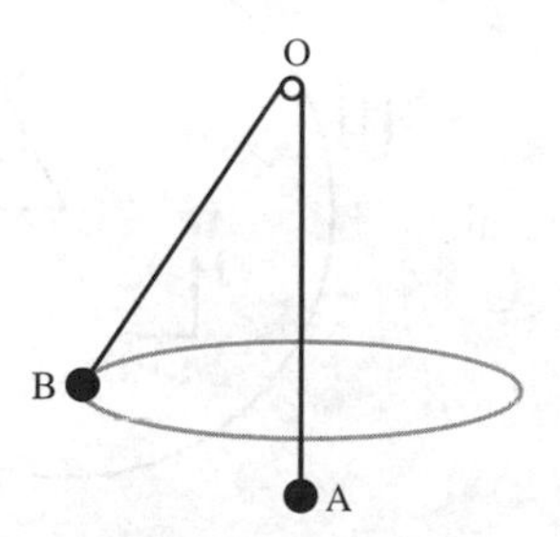

If the speed of B is 7 m s^{-1}, how long is the section BO of the string?

8 A pendulum bob, P, of mass 1.2 kg, hangs at one end of a light, inextensible string which passes through a smooth hole in a table at a point O. The length of OP is 0.7 m. The other end of the string is attached to a particle, Q, of mass 5.2 kg, which is resting on the rough horizontal surface of the table. The coefficient of friction between Q and the table is 0.25. The bob, P, is made to move as a conical pendulum below O. Find the maximum angular speed at which it can move without making Q slip.

9 A smooth wire has a bead of mass 0.005 kg threaded onto it. It is then bent round to form a circular hoop of radius 0.2 m. The hoop is fastened in a horizontal position, whilst the bead travels round it at a constant speed of 1 m s^{-1}. Find the magnitude and direction of the reaction force between the hoop and the bead.

10 The device illustrated is used as a governor for a steam engine. AB is a shaft which is turned by the engine. C is a collar which can slide on the shaft AB and which, in its lowest position, holds down a valve controlling the flow of steam to the piston. Spheres P and Q rotate with the shaft.

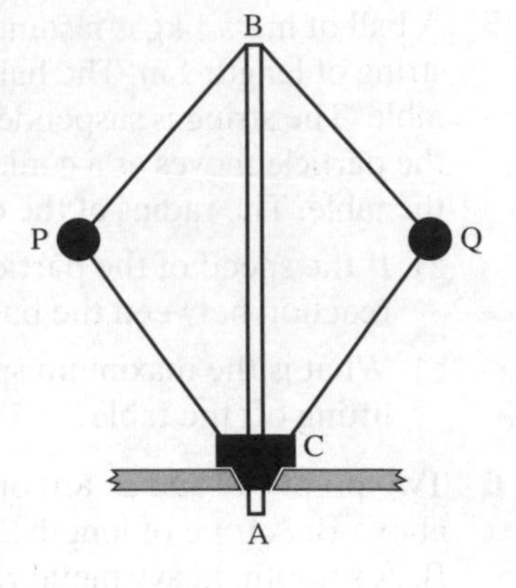

As the rate of rotation increases, so does the tension in the rods connecting the spheres to the point B and to the collar C. Eventually the speed is great enough for the tensions in the lower rods to lift the collar, reducing the flow of steam.

The spheres each have mass 3 kg and the collar has mass 6 kg. PBQC is a rhombus of side length 0.5 m. Angle PBC is 30°. Assuming that the rods are light and inextensible and that the spheres can be modelled as particles:

a) Find the tension in the lower rods in terms of ω, the angular speed of the shaft.

b) Find the value of ω if the collar is on the point of moving up the shaft.

M2

6.6 Circular motion with non-uniform speed

So far, you have considered only motion with uniform speed. You can now examine the more general situation.

Suppose the particle, P, moves in a circle so that at time t its position vector is

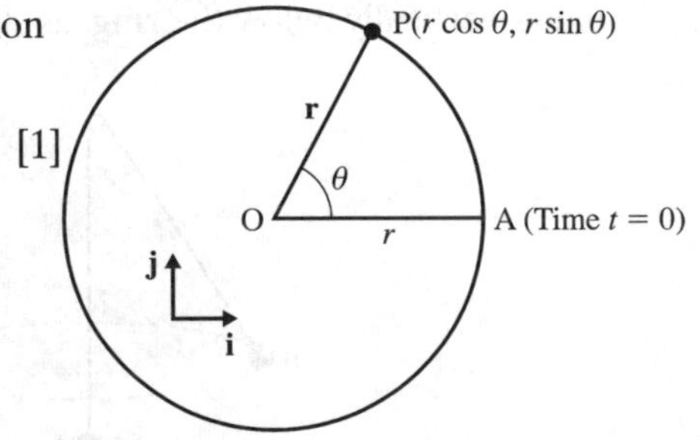

$$\mathbf{r} = r(\cos\theta\mathbf{i} + \sin\theta\mathbf{j}) \qquad [1]$$

as shown, where θ is a function of t and $\frac{d\theta}{dt} = \dot{\theta}$ is not, in general, constant.

By the chain rule,

$$\frac{d(\cos\theta)}{dt} = \frac{d(\cos\theta)}{d\theta} \times \frac{d\theta}{dt} = (-\sin\theta)\dot{\theta}$$

and $$\frac{d(\sin\theta)}{dt} = \frac{d(\sin\theta)}{d\theta} \times \frac{d\theta}{dt} = (\cos\theta)\dot{\theta}$$

Of course, r is a constant, so when you differentiate [1] to obtain the velocity vector, $\mathbf{v}$, you get

$$\mathbf{v} = r\left(\frac{d(\cos\theta)}{dt}\mathbf{i} + \frac{d(\sin\theta)}{dt}\mathbf{j}\right)$$

and so $$\mathbf{v} = r\dot{\theta}(-\sin\theta\mathbf{i} + \cos\theta\mathbf{j}) \qquad [2]$$

From [2] you get the fact that the speed of the particle is

$$|\mathbf{v}| = r\dot{\theta}.$$

This corresponds to $v = r\omega$, which you used for circular motion with uniform speed, but here $r\dot{\theta}$ is not constant.

You now differentiate [2] to obtain the acceleration vector, **a**. You need to use the product rule:

$$\mathbf{a} = r\frac{d\dot{\theta}}{dt}(-\sin\theta\mathbf{i} + \cos\theta\mathbf{j}) + r\dot{\theta}\left(-\frac{d(\sin\theta)}{dt}\mathbf{i} + \frac{d(\cos\theta)}{dt}\mathbf{j}\right)$$
$$= r\ddot{\theta}(-\sin\theta\mathbf{i} + \cos\theta\mathbf{j}) - r\dot{\theta}^2(\cos\theta\mathbf{i} + \sin\theta\mathbf{j})$$

Now, $(\cos\theta\mathbf{i} + \sin\theta\mathbf{j})$ is a unit vector in the direction of **r**. So, you can write

$$(\cos\theta\mathbf{i} + \sin\theta\mathbf{j}) = \hat{\mathbf{r}}$$

Similarly, $(-\sin\theta\mathbf{i} + \cos\theta\mathbf{j})$ is a unit vector in the direction of **v**, and so is tangential to the circle. You can write $(-\sin\theta\mathbf{i} + \cos\theta\mathbf{j}) = \hat{\mathbf{t}}$
You can then write the acceleration vector as

$$\mathbf{a} = r\ddot{\theta}\hat{\mathbf{t}} - r\dot{\theta}^2\hat{\mathbf{r}} \qquad [3]$$

From [3] you get two important facts:

- The particle has a **tangential component** of acceleration of $r\ddot{\theta}$.
- The particle has a **radial component** of acceleration of $-r\dot{\theta}^2$. This is directed towards the centre of the circle and corresponds to the $r\omega^2$ which you used for circular motion with uniform speed, but here $\dot{\theta}^2$ is not constant.

Motion in a vertical circle

A common example of circular motion with non-uniform speed is when a body is moving in a vertical circle. This occurs in many circumstances. Typically, a body may be rotating on the end of a string or rod, may be threaded onto a hoop or may be sliding on the inner or outer surface of a circular object. In all cases, the path may be a complete circle, or just a circular arc.

The situations listed can be divided into two categories. In the first, the body cannot leave the circular path. In the second, the body may leave the circular path at some stage of its motion.

The body cannot leave the circle
This includes bodies rotating on the end of rigid rods and beads threaded onto hoops. Such systems can behave in one of two ways:

- If the energy of the system is sufficient, the body rotates in a complete circle.
- If the energy is insufficient, the body cannot reach the highest point of the circle, but oscillates between two symmetrical positions, at each of which its speed is instantaneously zero, as shown in Figure 1.

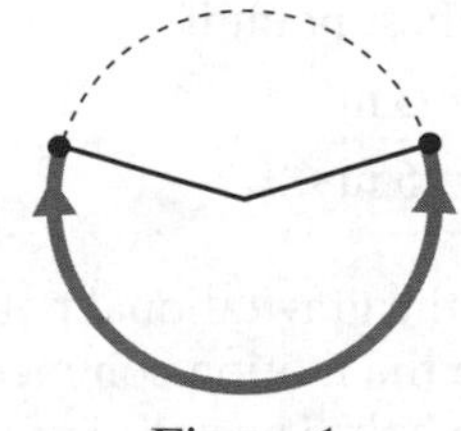

Figure 1

The body can leave the circle
This includes bodies rotating on the end of strings, or sliding on the surface of circular objects. Except for the case of a body sliding on the outer surface of a circular object, which is bound to leave the circle at some point, such systems can behave in any of three ways.

M2

- ✦ If the energy of the system is sufficient, the body rotates in a complete circle.
- ✦ If the energy of the system is so low that the body cannot rise beyond the level of the centre of the circle, it oscillates between two symmetrical positions, at each of which its speed is instantaneously zero, as shown in Figure 2.
- ✦ If the energy of the system is such that the body can rise above the level of the centre of the circle without being enough to carry it completely round the circle, it will leave the circle at some point and its motion will become that of a projectile, as shown in Figure 3.

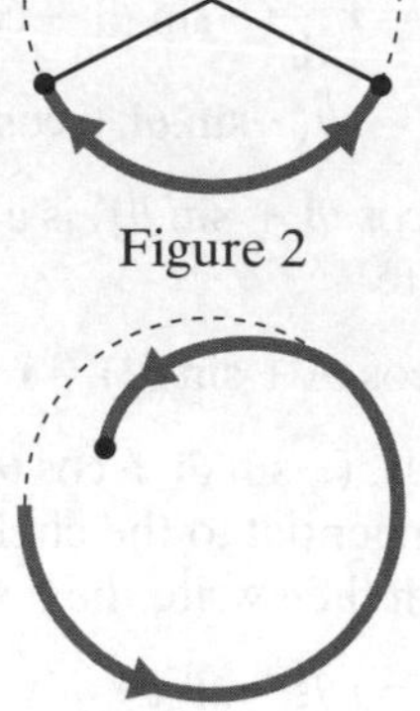

Figure 2

Figure 3

The mechanics of motion in a vertical circle

When modelling motion in a vertical circle, it is usual to make the following assumptions:

- ✦ The body is a particle.
- ✦ There is no air resistance.
- ✦ There is no loss of energy through any other resistance.
- ✦ If the particle is attached to a string or rod, this is light and inextensible.
- ✦ The path is a perfect circle. This assumption is often made when modelling situations such as roller-coaster loop-the-loops, in which the entry and exit paths do not coincide, but where the car travels in something very close to a circle.

In most problems concerning motion in a vertical circle, you will make use of the principle of conservation of energy.

Example 9

A pendulum consists of a bob, P, of mass 2 kg, attached to a light rod of length 1 m, the other end of which is freely hinged to a fixed point, O. The bob rests vertically below O, and is then given an impulse so that it starts moving with speed u. Find the angle between the rod and the downward vertical when the bob reaches its highest point if

a) $u = 3\text{ m s}^{-1}$

b) $u = 5\text{ m s}^{-1}$

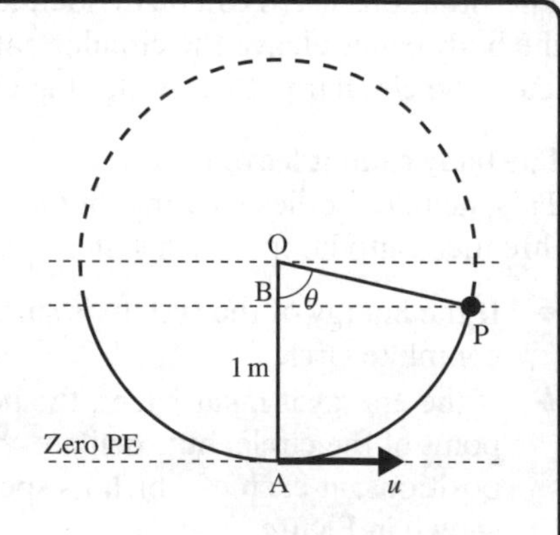

Take the gravitational potential energy to be zero at the point A, where the motion commences. Let P be the highest point reached by the bob. The speed of the bob at P is therefore zero.

OP = 1 m, so OB = $\cos\theta$ m. Hence, AB = $(1 - \cos\theta)$ m.

When the bob is at A, you have

$$\text{KE} = \tfrac{1}{2} \times 2 \times u^2 = u^2$$

$$\text{GPE} = 0$$

When the bob is its highest point, P, you have

$$KE = 0$$
$$GPE = 2g \times AB = 19.6(1 - \cos\theta)$$

Energy is conserved, so you have

$$u^2 = 19.6(1 - \cos\theta)$$

and so $\cos\theta = 1 - \dfrac{u^2}{19.6}$

a) When $u = 3$, you have

$$\cos\theta = 1 - \frac{9}{19.6} = 0.541 \quad \text{which gives} \quad \theta = 57.3°$$

b) When $u = 5$, you have

$$\cos\theta = 1 - \frac{25}{19.6} = -0.276 \quad \text{which gives} \quad \theta = 106°$$

In part b), the bob is above the level of O when it comes to rest. This is only possible when the pendulum involves a rod. Had string been used, the bob would have left the circle.

M2

Example 10

A hollow cylinder, of internal radius 2 m, rests with its axis horizontal. O is a point on that axis. A particle, P, of mass 5 kg, rests inside the cylinder at A vertically below O. The particle is given an impulse so that it starts to move on the smooth surface of the cylinder in a circle about O. Its initial speed is 8 m s^{-1}.

a) What is the normal reaction between the cylinder and the particle when the line OP makes an angle of 60° with the downward vertical?

b) What angle does OP make with the downward vertical when the normal reaction is zero?

c) What is the speed of the particle when the normal reaction is zero?

First, you need to use conservation of energy to find the speed of the particle when OP makes an angle θ with the downward vertical.

OP is 2 m, so OB = $2\cos\theta$. Hence,
AB = $2 - 2\cos\theta = 2(1 - \cos\theta)$ m

When the particle is at A, you have

$$KE = \tfrac{1}{2} \times 5 \times 8^2 = 160\text{ J}$$
$$GPE = 0$$

When the particle is at a general point, P, you have

$$KE = \tfrac{1}{2} \times 5 \times v^2 = 2.5v^2$$
$$GPE = 5g \times AB = 98(1 - \cos\theta)$$

Energy is conserved, so you have

$$2.5v^2 + 98(1 - \cos\theta) = 160$$

and so $v^2 = 8(3.1 + 4.9\cos\theta)$ [1]

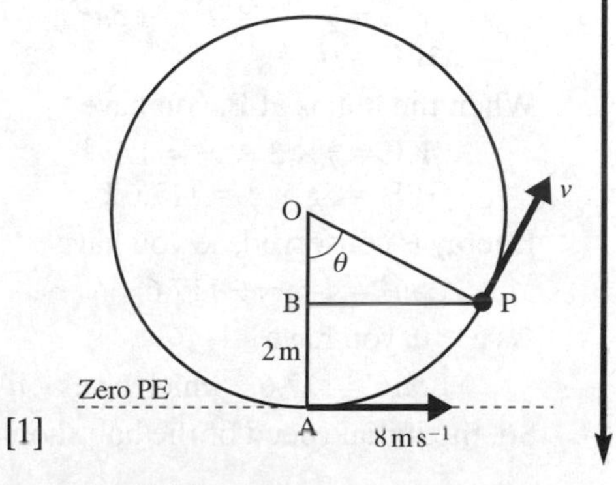

You now need to consider the forces acting on the particle. Because the surface is smooth, the only forces acting on the particle are its weight, 49 N, and the normal reaction, R.

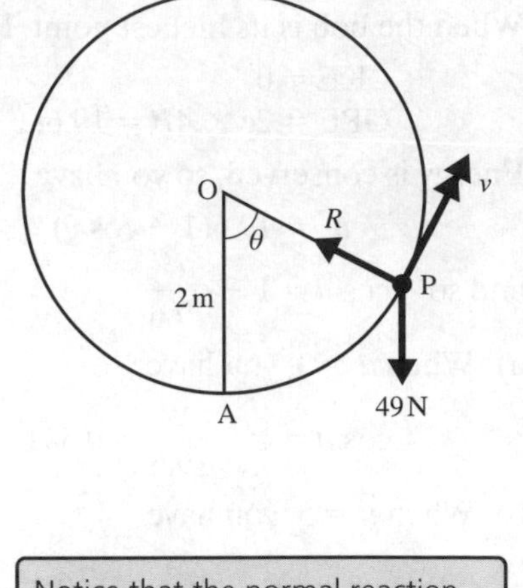

The component of acceleration in the direction PO is $\frac{v^2}{2}$.

So, resolving in the direction PO, you have

$$R - 49\cos\theta = 5 \times \frac{v^2}{2}$$

which gives $R = 2.5v^2 + 49\cos\theta$

Substituting from [1], you have

$$R = 20(3.1 + 4.9\cos\theta) + 49\cos\theta$$

which gives $R = 62 + 147\cos\theta$ [2]

a) When $\theta = 60°$, you have from [2]

$$R = 62 + 147 \times \tfrac{1}{2} = 135.5\text{ N}$$

b) When $R = 0$, you have from [2]

$$62 + 147\cos\theta = 0$$

which gives $\cos\theta = -0.422$ and so $\theta = 115°$

c) Substituting $\cos\theta = -0.422$ into [1], you get

$$v^2 = 8(3.1 + 4.9 \times (-0.422)) = 8.27$$

and so $v = \sqrt{8.27} = 2.86\text{ m s}^{-1}$

Notice that the normal reaction becomes zero at the point where the particle leaves the circle. At this point, it is travelling with a speed of 2.86 m s^{-1} at an angle of 65° to the horizontal. Its subsequent motion is as a projectile with this initial velocity.

M2

Example 11

A pendulum consists of a bob, P, of mass 3 kg, attached to the end of a light rod of length 2 m, whose other end is freely hinged at the point O. The bob rests at its lowest point. It is then given an impulse and commences to move at speed u m s^{-1}.

a) Find the minimum value of u, given that the pendulum performs a complete circle.
b) If the rod were replaced by a string, find the minimum value of u for the pendulum to perform a complete circle.

a) Suppose the pendulum reaches the top, B, with speed $v \geqslant 0$. Take the zero level for potential energy as shown.

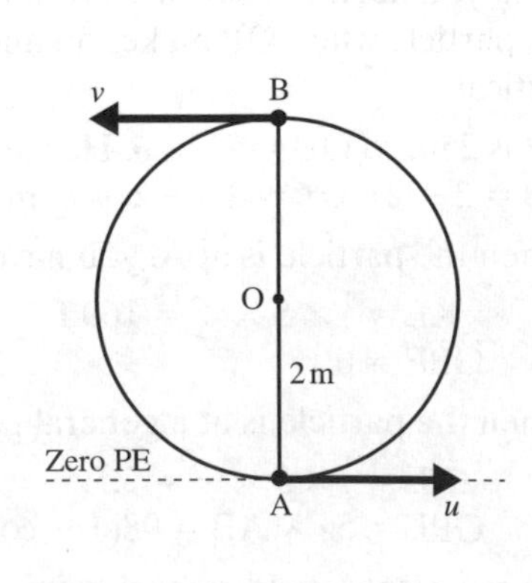

When the bob is at A, you have

$$\text{KE} = \tfrac{1}{2} \times 3 \times u^2 = 1.5u^2$$
$$\text{GPE} = 0$$

When the bob is at B, you have

$$\text{KE} = \tfrac{1}{2} \times 3 \times v^2 = 1.5v^2$$
$$\text{GPE} = 3g \times 4 = 117.6\text{ J}$$

Energy is conserved, so you have

$$1.5u^2 = 1.5v^2 + 117.6$$ [1]

As $v \geqslant 0$, you have

$$1.5u^2 \geqslant 117.6 \quad \text{which gives} \quad u \geqslant 8.85$$

So, the initial speed of the bob should be at least 8.85 m s^{-1}.

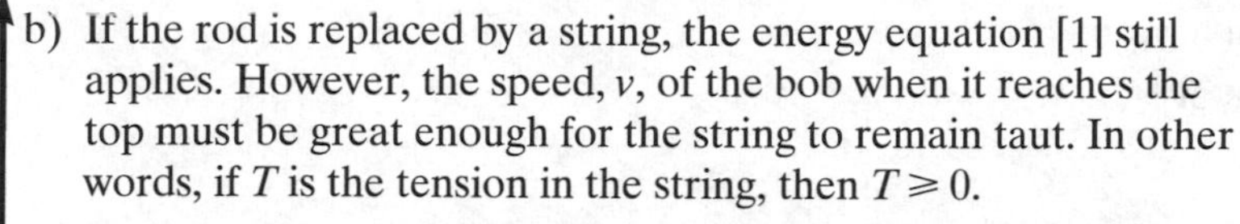

b) If the rod is replaced by a string, the energy equation [1] still applies. However, the speed, v, of the bob when it reaches the top must be great enough for the string to remain taut. In other words, if T is the tension in the string, then $T \geqslant 0$.

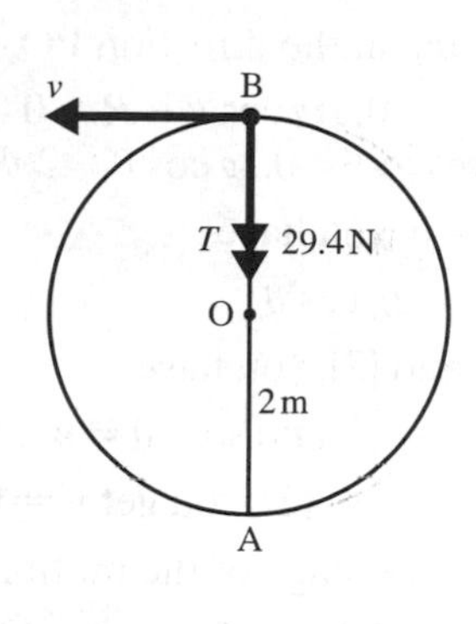

At B, the acceleration of the bob towards the centre of the circle is $\frac{v^2}{2}$. So, resolving in the direction BO, you have

$$T + 29.4 = 3 \times \frac{v^2}{2}$$

> $mg = 3 \times 9.8$
> $= 29.4\,\text{N}$

As $T \geqslant 0$, you have

$$\frac{3v^2}{2} \geqslant 29.4 \quad \text{and so} \quad v^2 \geqslant 19.6 \qquad [2]$$

Substituting from [2] into [1], you get

$$1.5u^2 \geqslant 1.5 \times 19.6 + 117.6$$

which gives $u \geqslant 9.90$

So, for a string, the initial speed of the bob should be at least $9.90\,\text{m s}^{-1}$.

M2

Example 12

A particle of mass 0.01 kg rests at the top of a smooth sphere of radius 0.2 m which is fixed to a horizontal table. It is displaced slightly so that it slides off the sphere. How far from the point of contact between the sphere and the table does the particle strike the table?

There are two stages to the motion of the particle. In the first stage, the particle slides down the outside of the sphere. At some point it leaves the surface, leading to the second stage in which the particle moves as a projectile. You need to find the position of the particle and its velocity at the point where it leaves the surface. This will provide the initial conditions for the projectile motion.

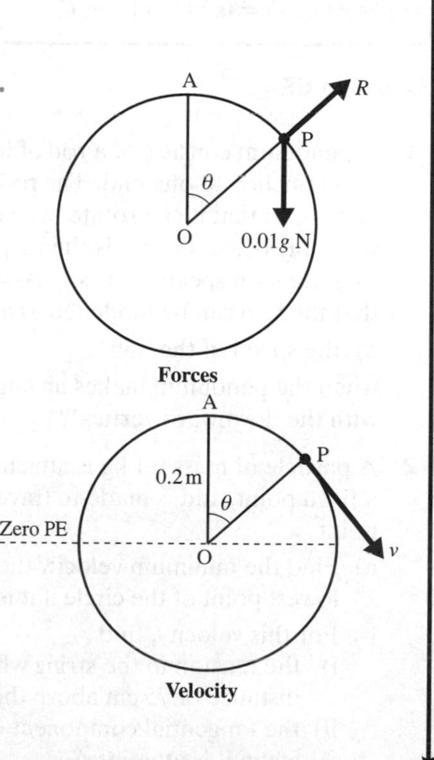

The particle will leave the surface at the point where the reaction between the particle and the surface becomes zero.

At point A, you have

$$\text{KE} = 0$$
$$\text{GPE} = 0.01 \times g \times 0.2 = 0.002g$$

At a general point, P, you have

$$\text{KE} = \tfrac{1}{2} \times 0.01 \times v^2 = 0.005v^2$$
$$\text{GPE} = 0.01 \times g \times 0.2 \cos\theta = 0.0002g \cos\theta$$

Energy is conserved, so you have

$$0.005v^2 + 0.002g\cos\theta = 0.002g$$

and so $v^2 = 0.4g(1 - \cos\theta) \qquad [1]$

The acceleration towards the centre of the circle is $\frac{v^2}{0.2} = 5v^2$.

So, resolving in the direction PO, you have

$$0.01g\cos\theta - R = 0.01 \times 5v^2$$

which gives $v^2 = 0.2g\cos\theta - 20R$

When $R = 0$, you have

$$v^2 = 0.2g\cos\theta \qquad [2]$$

From [1] and [2], you have

$$\cos\theta = \tfrac{2}{3} \quad \text{and so} \quad \theta = 48.2°$$

Substituting into [2], you get $v = 1.14\ \text{m s}^{-1}$

The projectile stage of the motion starts as shown in the diagram.

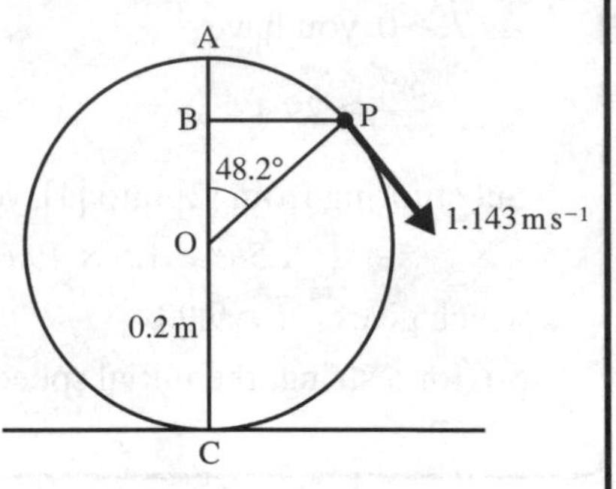

The distance $BC = 0.2(1 + \cos 48.2°) = \tfrac{1}{3}$ m

The vertical component of the initial velocity is $1.14\sin 48.2° = 0.852\ \text{m s}^{-1}$.

Using $s = ut + \tfrac{1}{2}at^2$, the particle reaches the table at time t, where

$$\tfrac{1}{3} = 0.852t + \tfrac{1}{2}gt^2$$

which gives $4.9t^2 + 0.852t - \tfrac{1}{3} = 0$

Solving this quadratic equation, we get $t = 0.188$ s (or -0.362 s).

The horizontal component of the initial velocity is $1.14\cos 48.2° = 0.762\ \text{m s}^{-1}$.

In 0.188 s, the particle travels $0.762 \times 0.188 = 0.143$ m horizontally from the point where it left the sphere.

As $BP = 0.2\sin 48.2° = 0.149$ m, the particle lands $0.149 + 0.143 = 0.292$ m from C.

M2

Exercise 6E

1 A pendulum consists of a rod of length 1 m with a bob of mass 2 kg attached to one end. The rod is freely pivoted at the other end, O, so that it can rotate in a vertical circle. Initially the bob is vertically below O, and is then given an impulse so that it starts to move with speed 6.5 m s^{-1}. Assuming that the rod is light and that the bob can be modelled as a particle, what is

a) the speed of the bob b) the force in the rod

when the pendulum makes an angle of i) 30° ii) 90° and iii) 150° with the downward vertical?

2 A particle of mass 0.1 kg is attached by a string of length 1.5 m to a fixed point, and is made to travel in a vertical circle about that point.

a) Find the minimum velocity the particle must have at the lowest point of the circle if it is to make complete revolutions.

b) For this velocity, find

i) the tension in the string when the particle is at point A, a distance of 75 cm above the lowest point

ii) the tangential component of the particle's acceleration when it is at point A.

3 A particle of mass 0.01 kg is placed on the topmost point, A, of a smooth sphere of centre O and radius 0.5 m. It is slightly displaced. When it reaches point B it is about to leave the surface of the sphere. Calculate the angle AOB.

4 A particle of mass m hangs at rest, suspended from a point, O, by a light, inextensible string of length a. The particle receives an impulse so that it starts moving with speed $\sqrt{3ga}$. Find the angle between the string and the vertical when it goes slack.

5 A stone of mass 0.5 kg performs complete revolutions in a vertical circle on the end of a light, inextensible string of length 1 m. Show that the string must be strong enough to support a tension of at least 29.4 N.

6 A particle of mass m travels in complete vertical circles on the end of a light, inextensible string of length a. If the maximum tension in the string is three times the minimum tension, find the speed of the particle as it passes through the lowest point on the circle.

7 A pendulum of length a has a bob of mass m. The speed of the bob at the lowest point of its path is U. Find the condition which U must satisfy for the bob to make complete revolutions if the pendulum consists of a) a rod b) a string.

8 A particle of mass m is projected horizontally with speed v from the topmost point, A, of a sphere of radius a and centre O. It remains in contact with the sphere until leaving the surface at point B. If angle AOB is 30°, find v.

9 A particle of mass 2 kg is attached to the end of a light, inextensible string of length 1 m, the other end of which is attached to a fixed point, O. The particle is held with the string taut and horizontal, and is released from rest. When the string reaches the vertical position, it meets a fixed pin, A, a distance x below O. Given that the particle just completes a circle about A, find the value of x.

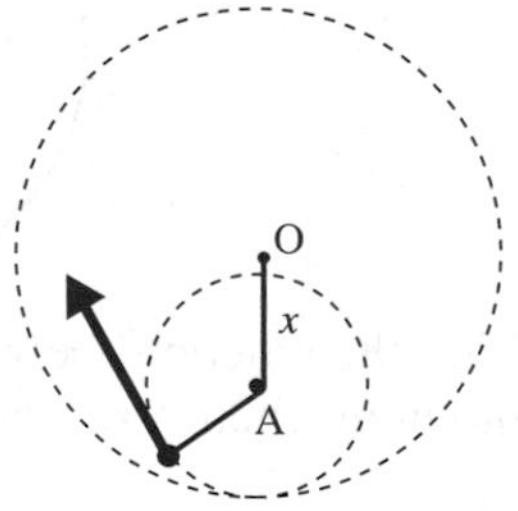

O
x
A

10 A bead of mass m is threaded onto a smooth, circular hoop of radius a, which is fixed in a vertical plane. The bead is displaced from rest at the top of the hoop. Find the **resultant** acceleration of the bead when it has reached a point which is a vertical distance $\frac{3}{4}a$ below its starting point.

11 A body of mass 40 kg is swinging on the end of a light rope of length 3 m, which is attached to a fixed point 3.6 m above horizontal ground. The body moves so that at the extreme positions of its motion, the rope makes an angle of 60° with the downward vertical through O. At an instant when the pendulum makes an angle of 30° with the downward vertical through O and the body is rising, the body breaks free from the rope. Calculate the horizontal displacement of the body from O at the point where it hits the ground.

12 A pendulum bob of mass m is fastened to one end of a string of length r whose other end is fixed at a point O. The bob is at rest in its lowest position when it is set in motion with initial speed $\sqrt{\frac{7gr}{2}}$. As it swings upwards, the string meets a small, fixed peg, P, on the same level as O. The string then wraps round P. What is the closest that P can be to O so that the bob makes a complete revolution about P?

13 A ring of mass 5kg is threaded onto a rope of length 10 m, whose ends are attached to two fixed points 6 m apart and on the same level. The ring hangs at rest. It is then set in motion so that it travels on a circular path whose plane is perpendicular to the line joining the two ends of the rope. Given that the ring can just make complete revolutions, find the maximum tension in the rope. State any modelling assumptions you have made in reaching your answer.

M2

14 Particles A and B, of masses m and $2m$ respectively, are connected by a light, inextensible string of length πa. The particles are placed symmetrically, and with the string taut, on the smooth outer surface of a cylinder of radius $3a$, as shown, and the system is released from rest.

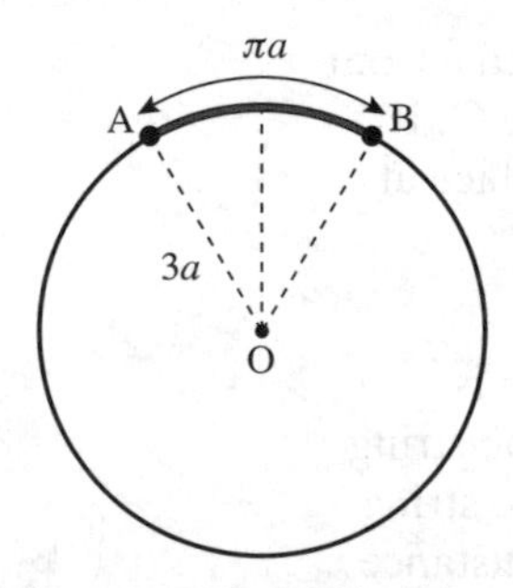

Find the reactions between the cylinder and the particles at the moment when A reaches the topmost point.

15 The diagram shows a loop-the-loop on a roller-coaster ride. The car approaches the loop on a horizontal track. The maximum speed at which the car can enter the loop is 80 km h^{-1}.

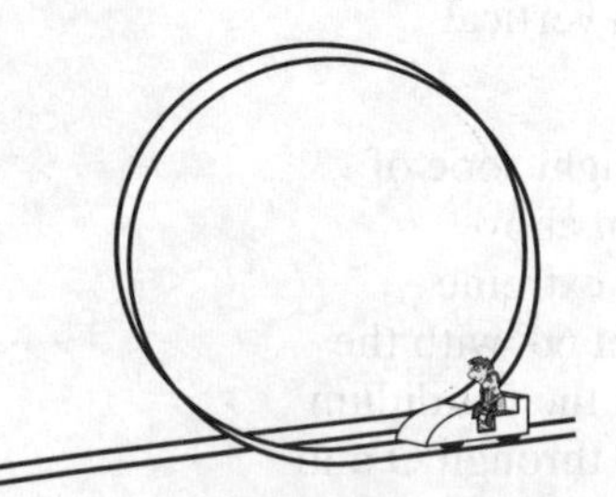

What is the greatest radius with which the loop can be constructed if the car is not to leave the track?

Summary

You should know how to ...	Check out
1 Convert between different units for angular speed.	**1** A body is rotating with angular speed 5 rad s^{-1}. Calculate its angular speed in rev min^{-1} (rpm).
2 Know and use the relationships $v = r\omega$, $a = r\omega^2$, $a = \frac{v^2}{r}$	**2** A particle is rotating in a circle with centre O. Its constant angular speed is 6 rad s^{-1}. It has speed 30 ms $^{-1}$. Find a) the distance of the particle from the axis of rotation b) the magnitude and direction of its acceleration.
3 Find the position, velocity and acceleration vectors of a particle moving in a circle.	**3** The position vector of a particle is $\mathbf{r} = 5\cos 3t\mathbf{i} + 5\sin 3t\mathbf{j} - 2\mathbf{k}$ a) Find i) $\dot{\mathbf{r}}$ ii) $\ddot{\mathbf{r}}$ b) State the centre of the circle around which the particle is moving.
4 Model the motion of a particle moving as a conical pendulum.	**4** A particle of mass 3 kg is attached to a light inextensible string of length 2 m. The other end of the string is attached to a point A. The particle is moving in a horizontal circle of radius 1 m below A. Find a) the tension in the string b) the angular speed of the particle.
5 Model the motion of a particle moving in a vertical circle.	**5** A particle of mass m is suspended by an inextensible string of length r from a fixed point O. The string remains taut as the particle rotates in a vertical circle about O. Its speed, v, at the highest point of its path is half its speed at the lowest point. Find a) v in terms of g and r b) the tension in the string when the particle is at the lowest point.

Revision exercise 6

1 A children's roundabout has a horizontal, circular base with centre O. Stephanie places a doll, D, on the roundabout and then pushes the roundabout so that it rotates with constant angular speed ω radians per second. Stephanie notices that the doll makes 8 complete revolutions about O in one minute.

a) Show that $\omega = \dfrac{4\pi}{15}$.

b) The doll's speed is $0.75\ \text{m s}^{-1}$. Find its distance from O, giving your answer to 2 significant figures.

c) Find the acceleration of the doll and indicate its direction on a diagram.

(*AQA, 2002*)

2 A game uses a light inextensible string, of length 50 cm, attached to one end B of a fixed vertical pole, AB, as shown in the diagram. A ball, C, of mass 0.1 kg is attached to the other end of the string. The ball is moving in a horizontal circle around the pole with constant angular speed. The string makes an angle θ with the pole, where $\cos\theta = 0.8$.

a) Draw a diagram showing the forces acting on the ball.

b) Show that the tension in the string is 1.225 N.

c) Determine the angular speed of the ball in radians per second.

(*AQA, 2001*)

3 A particle, P, of mass 2 kg is attached to two strings of lengths 0.2 m and 0.1 m. The strings are fixed to the particle and to the points A and B respectively. The point A is directly above the point B. The particle describes a horizontal circle, centre B and radius 0.1 m, at a speed of $4\ \text{m s}^{-1}$. The particle and strings are shown in the diagram.

a) Calculate the magnitude of the acceleration of the particle.

b) Find the tension in the upper string.

c) Find the tension in the lower string. (*AQA, 2003*)

4 A particle of mass 0.4 kg is attached at the point P to two light strings, QP and OP. The points O and Q are fixed with Q at a distance of 0.4 m vertically above O. The string QP is inextensible and of length 0.5 m. The string OP is elastic and of natural length 0.2 m and stiffness $k\ \text{N m}^{-1}$. The particle moves in a horizontal circle, centre O and radius 0.3 m, at a constant angular speed of $5\ \text{rad s}^{-1}$.

a) Draw a diagram showing the forces acting on the particle.

b) Show that the tension in the string QP is 4.9 N.

c) Write down, in terms of k, the tension in the string OP.

d) Show that $k = 0.6$.

e) Find the elastic potential energy stored in the string OP.

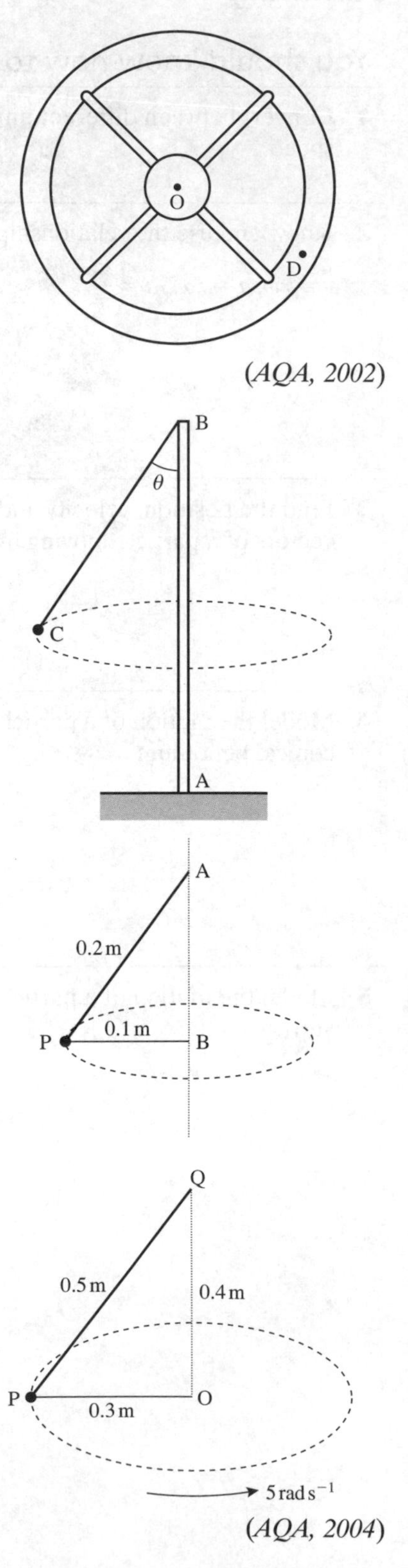

(*AQA, 2004*)

5 A conical pendulum consists of a small sphere of mass 3 kg attached to the end of a light inextensible string of length 0.6 m. The sphere moves in a horizontal circle at a constant speed.

Model the sphere as a particle.

a) The string is inclined at an angle of 30° to the vertical, as shown in the diagram.
   i) Find the tension in the string.
   ii) Show that the angular speed of the sphere is 4.34 radians per second, to three significant figures.

b) The angular speed of the particle is now doubled. Find the new angle between the string and the vertical, giving your answer to the nearest degree.

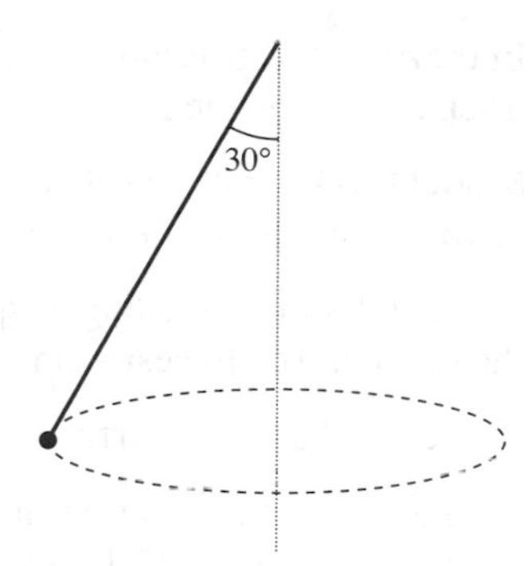

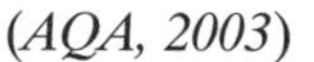
(*AQA, 2003*)

6 A bead, of mass m, is threaded onto a smooth circular ring, of radius r, which is fixed in a vertical plane. The bead is moving on the wire. Its speed, v, at the highest point of its path is one quarter of its speed at the lowest point.

a) Show that $v = \sqrt{\frac{4}{15}gr}$.

b) Find the reaction of the wire on the bead, in terms of m and g, when the bead is:
   i) at the highest point;
   ii) $\frac{1}{2}r$ below its highest point.

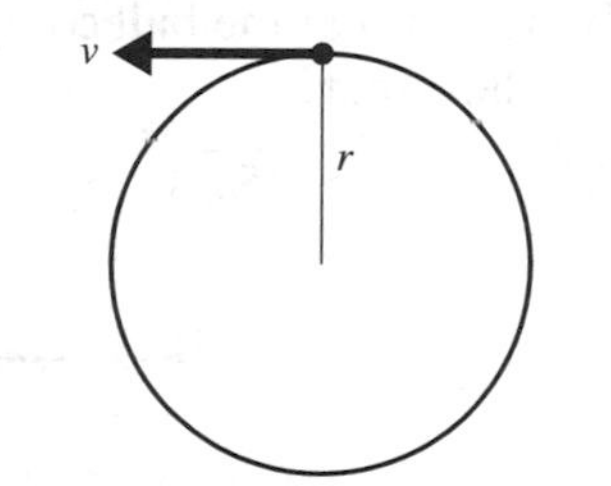

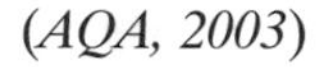
(*AQA, 2003*)

7 One end of a light inextensible string of length a is attached to a fixed point, O, and a particle of mass m is attached to the other end, A. The particle is held so that the string is taut and OA is horizontal. It is then projected vertically downwards with speed u as shown in the diagram.

The string becomes slack when OA is inclined at an angle of 60° above the horizontal.

a) Show that the speed of the particle when the string becomes slack is $\sqrt{\frac{\sqrt{3}}{2}ag}$.

b) Hence find u in terms of a and g.

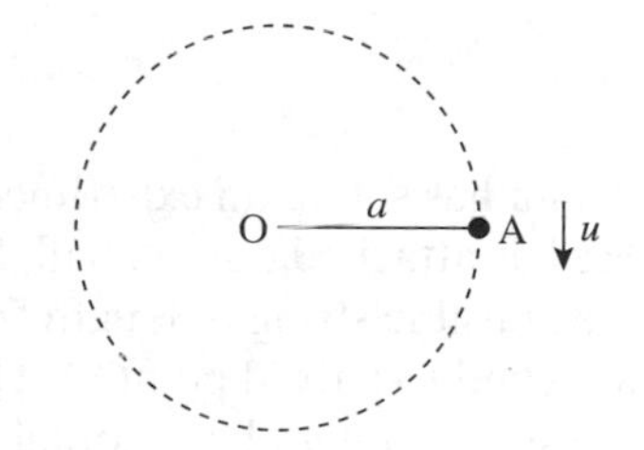

8 A smooth hemisphere of radius l and centre Q lies with its plane face fixed to a horizontal surface. A particle, P, of mass m can move freely on the surface of the hemisphere.

The particle is set in motion along the surface of the hemisphere with a speed, u, at the highest point of the hemisphere.

a) Show that, while the particle is in contact with the hemisphere, the velocity of the particle when PQ makes an angle θ to the vertical, is

$$(u^2 + 2gl[1 - \cos\theta])^{\frac{1}{2}}.$$

b) Find, in terms of l, u and g, the cosine of the angle θ when the particle leaves the surface of the hemisphere.

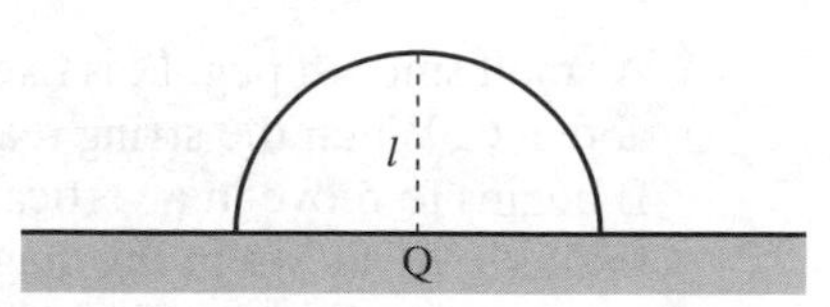

(*AQA, 2002*)

M2

9 In crazy golf, a golf ball is fired along a smooth track and loops the loop inside a section of track.

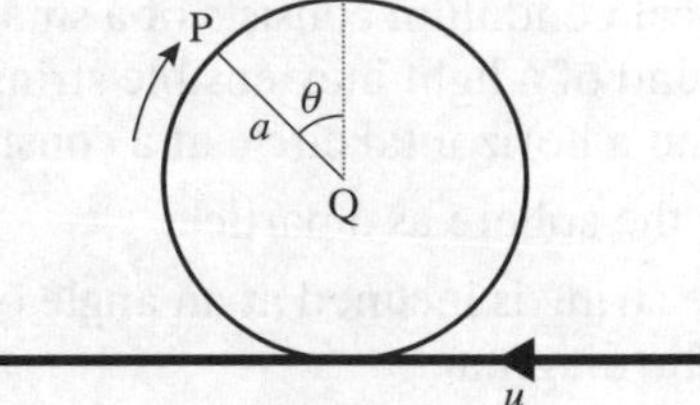

Model this loop as a vertical circle of radius a and centre Q, as shown in the diagram.

The golf ball is travelling at speed u as it enters the circle at the lowest point.

Model the ball as a particle P, of mass m.

a) Show that the reaction of the track on the particle when QP makes an angle of θ with the upward vertical is

$$\frac{mu^2}{a} - 3mg\cos\theta - 2mg$$

b) Given that the ball completes a vertical circle inside the track, show that

$$u \geqslant \sqrt{(5ag)}$$

(AQA, 2004)

M2

10

B 5a O

5a

C

Adam has set up an experiment for his Mechanics coursework. He has attached a small ball, B, of mass m, to one end of a light inextensible string of length $5a$. The other end of the string is attached to a fixed point O. The ball is released from rest with the string taut and horizontal, as shown in the diagram. The ball subsequently passes through the point C, which is a vertical distance $5a$ below O.

a) Find an expression, in terms of a and g, for the speed of B when it reaches C.

b) A small smooth peg, P, is fixed at a distance d vertically above C. When the string reaches the vertical position, B begins to move in a vertical circle with centre P and radius d, as shown in the diagram.

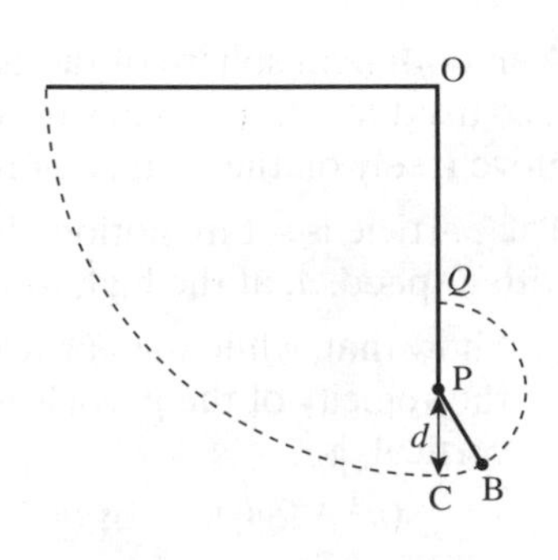

The ball reaches Q, the point at a distance d vertically above P, with speed v. At Q, the string is taut.

i) Show that $v^2 = 2g(5a - 2d)$.

ii) Find, in terms of a, d, g and m, the tension in the string when the ball is at Q.

iii) Hence show that $d < 2a$.

c) State **one** modelling assumption used in this question.

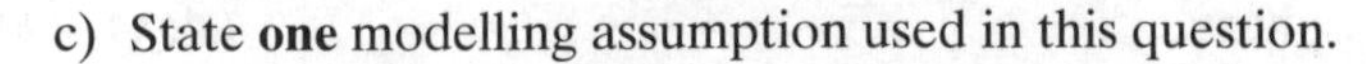

(AQA, 2004)

11 The diagram shows a particle P, of mass m, which is attached by a light inextensible string, of length l, to a fixed support O. The particle moves in a vertical plane. It is initially set in motion with speed u at right angles to OP, from the position where OP makes an angle α with the downward vertical through O.

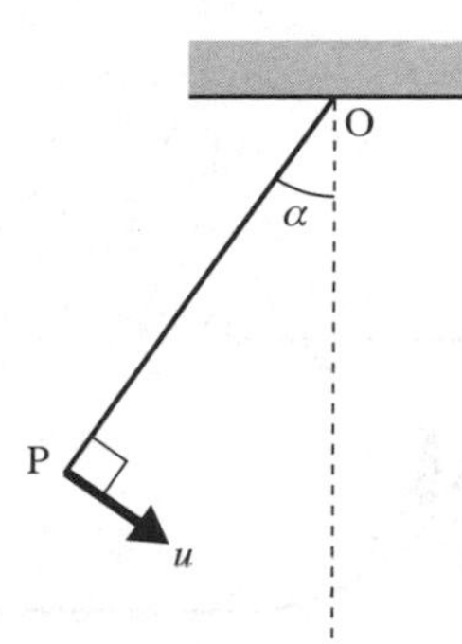

a) Show that when OP makes an angle θ with the downward vertical through O the speed, v, of the particle is given by

$$v^2 = u^2 + 2gl(\cos\theta - \cos\alpha).$$

b) At an adventure playground a girl of mass 40 kg swings on the end of a rope of length 5 m. The motion is in a vertical plane. Initially the rope makes an angle of 30° with the downward vertical and the girl has a speed of 2 m s^{-1} at right angles to the rope.

i) Show that the maximum speed of the girl during the motion is approximately 4.1 m s^{-1}.

ii) Determine the maximum angle that the rope makes with the downward vertical.

iii) Find the maximum tension in the rope.

iv) State one modelling assumption which you have used in this problem.

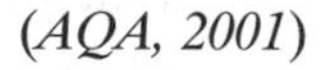

(*AQA, 2001*)

12

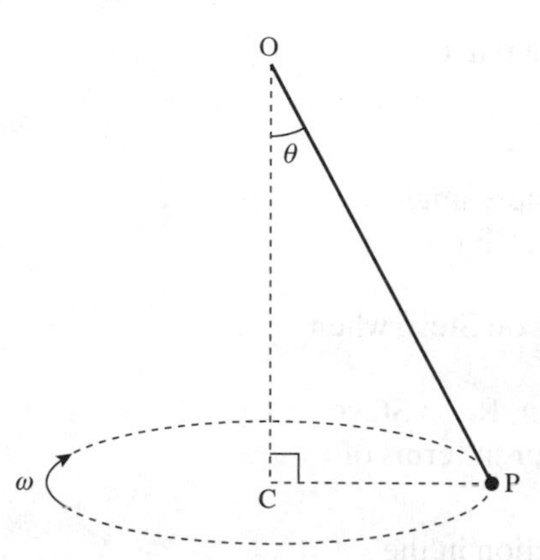

A particle, P, of mass m is attached by a light, inextensible thread of length l to a fixed point O. The particle moves with constant angular speed ω in a horizontal circle with centre C. The point C is vertically below O. The thread is inclined to the vertical at an angle θ, as shown in the diagram.

a) Find an expression for the tension in the thread. Give your answer in terms of m, l and ω.

b) Show that $\cos\theta = \dfrac{g}{l\omega^2}$.

c) The greatest tension that the thread can withstand without breaking is 16 N.

In the case when $l = 0.4$ m and $m = 0.1$ kg:

i) show that $\omega \leqslant 20$ radians per second;

ii) find the greatest possible value of θ, giving your answer to the nearest degree.

(*AQA, 2003*)

13 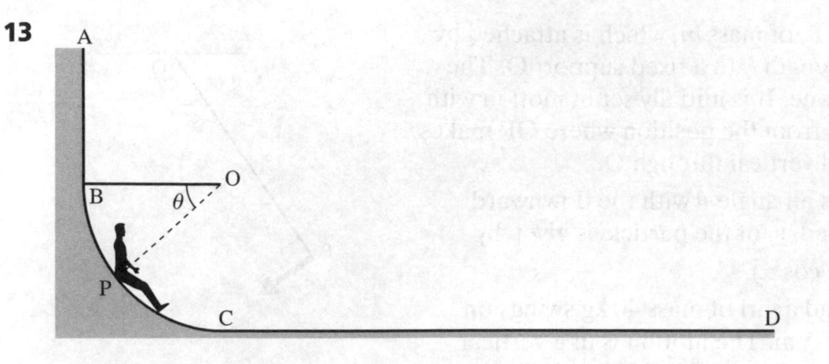

The diagram shows a vertical cross section of a new adventure slide at a theme park. It consists of three sections AB, BC and CD.
Section AB is smooth and vertical and has length r.
Section BC is smooth and forms a quarter of a circle. This circle has centre O and radius r. The radius OB is horizontal and OC is vertical.
Section CD is rough, straight and horizontal. It is of length $4r$.
Steve, who has mass m, starts from rest at A and reaches speed u at the point B. He remains in contact with the surface until he reaches D.

It can be assumed that Steve can be modelled as a particle throughout the motion.

a) Find u^2 in terms of g and r.

b) Steve reaches the point P between B and C where angle POB $= \theta$, as shown in the diagram. His speed at P is v.
   i) Show that $v^2 = 2gr(1 + \sin\theta)$
   ii) Draw a diagram showing the forces acting on Steve when he is at the point P.
   iii) Find an expression for the normal reaction, R, on Steve when he is at the point P. Give your answer in terms of m, g and θ.

c) Show that, as Steve crosses C, there is a reduction in the normal reaction of magnitude $4\,mg$.

d) Between C and D, Steve decelerates uniformly and comes to rest at the point D.
   Find his retardation. *(AQA, 2003)*

M2

7 Differential equations

This chapter will show you how to

- ✦ Use $v = \dfrac{dx}{dt}$ and $a = \dfrac{dv}{dt} = \dfrac{d^2x}{dt^2}$ to form differential equations.
- ✦ Solve differential equations to obtain relationships between velocity and time.
- ✦ Solve differential equations to obtain relationships between displacement and time.

Before you start

You should know how to ...	Check in
1 Rearrange equations.	**1** Rearrange $mp - mg = p$ to find p.
2 Integrate polynomials and algebraic, exponential, logarithmic and trigonometric functions.	**2** Find a) $\int (t^4 + 2t^3)\,dt$ b) $\int \left(\sqrt{t} + \dfrac{3}{t^2}\right) dt$ c) $\int e^{7t}\,dt$ d) $\int 2\sin 4t\,dt$
3 Integrate expressions, by means of substitution, partial fractions and integration by parts.	**3** Find a) $\int t(t-2)^3\,dt$ b) $\int \dfrac{2t+14}{(t-2)(t+4)}\,dt$ c) $\int te^{-3t}\,dt$ d) $\int \ln t\,dt$

M2

In Chapter 1 you met some examples of differential equations. This chapter presents a more extensive and detailed treatment.

7.1 Definitions and classification

> A differential equation is any equation containing a derivative.

Here are four examples:

i) $\dfrac{dy}{dx} + 2y = \sin x$ ii) $3\dfrac{d^2x}{dt^2} - 3\dfrac{dx}{dt} + 6x = 4\cos 3t$

iii) $\left(\dfrac{dy}{dt}\right)^2 - y\dfrac{d^2y}{dt^2} = \dfrac{dy}{dt}$ iv) $\dfrac{dy}{dx} - x^2y = x^3 - 3x + 4$

'Solving' a differential equation means obtaining a relationship connecting the two variables which does not involve a derivative.

A derivative corresponds to a rate of change. Many real-life problems, in mechanics and a variety of other fields, involve quantities which are changing continuously, and the associated mathematical models will be expressed in the form of differential equations. The solution of such equations is therefore an important branch of mathematics. You have already met simple differential equations such as $\frac{dv}{dt} = 2t$, which you solved directly by integration. Some other types of differential equation can also be solved using analytical techniques, but many other differential equations can only be solved by numerical methods.

Differential equations are categorised by their style and complexity.

- The **order** of a differential equation is the **order of the highest derivative** appearing in the equation. Of the examples above, **i** and **iv** are first order equations, while **ii** and **iii** are second order equations.
- The **degree** of a differential equation in y is the **highest power of y or its derivatives** appearing in the equation. Of the above examples, **i, ii** and **iv** are first degree equations, while **iii** is a second degree equation because it contains $\left(\frac{dy}{dt}\right)^2$. An equation of first degree is called **linear.**

M2

In this chapter you will only meet linear equations.

7.2 Forming differential equations

Before learning how to solve differential equations, you should consider situations in which they might arise.

Any situation involving a continuous rate of change can be modelled using differential equations. In mechanics the rate of change is usually with respect to time, and if this is the case it is not necessary to say so explicitly – you just refer to the **rate of change.** If the independent variable is not time, you have to be explicit. For example, in a structure made of heavy beams you might be interested in the rate at which the stresses in the beam change as you move along the beam, so you would use quantities such as $\frac{dS}{dx}$, which is the **rate of change of shearing force with respect to displacement.**

The majority of differential equations you will encounter in mechanics will arise as a result of applying Newton's second law.

Example 1

An object of mass 2 kg is travelling at a constant 20 m s^{-1} when it encounters a resistance which is proportional to its speed. Given that its initial acceleration is -5 m s^{-2}, write a differential equation to express the situation.

Let the resistance force be R.
You are given that $R \propto v$, and hence you have $R = kv$.

From Newton's second law ($F = ma$), you use $a = \frac{dv}{dt}$ to find

$$2\frac{dv}{dt} = -kv$$

You are given that $\frac{dv}{dt} = -5$ when $v = 20$. Hence you have

$$2 \times (-5) = -k \times 20$$

which gives $k = 0.5$

The required differential equation is therefore

$$2\frac{dv}{dt} = -0.5v$$

or

$$\frac{dv}{dt} = -0.25v$$

Example 2

A particle of mass 5 kg rests on a smooth horizontal plane. It is attached to a fixed point O on the plane by means of a light elastic string of natural length 2 m and modulus of elasticity 20 N. The particle is moved to a point 5 m from O and released from rest. Write a differential equation connecting x m, the distance the particle has moved and t s, the elapsed time.

When the particle has moved a distance x, the extension of the string is $(3 - x)$.

From Hooke's law, the force on the particle is

$$\frac{20(3-x)}{2} = 10(3-x)\text{ N.}$$

Hence, from Newton's second law, you use $a = \frac{d^2x}{dt^2}$ to find

$$5\frac{d^2x}{dt^2} = 10(3-x)$$

and hence $\frac{d^2x}{dt^2} = 2(3-x)$, which is the required differential equation.

Exercise 7A

1 An object of mass 5 kg is dropped from a hot air balloon. As it falls, it is subjected to air resistance R N proportional to its speed v m s^{-1}. By considering Newton's second law, express this as a differential equation connecting v, the time t s and a constant k. As it falls, the object's acceleration decreases and it approaches terminal velocity. If the terminal velocity of the object is 60 m s^{-1}, find the value of k.

2 A particle moves in a straight line so that its acceleration at time t seconds is proportional to $\sin \pi t$. Express this as a differential equation connecting its velocity v m s^{-1}, the time t s and a constant k. The maximum acceleration of the particle is 4 m s^{-2}. State the value of k, and the smallest positive value of t for which the maximum acceleration occurs.

3 A vehicle of mass 500 kg is driven along a straight horizontal road. The engine has power 5 kW and the resistance to motion is proportional to v^2, where v m s^{-1} is the velocity of the vehicle at time t s. Write the equation of motion of the vehicle as a differential equation connecting v, t and a constant k.

4 A particle of mass m kg moves along a straight line. At time t s, its displacement from a fixed point, O, on the line is x m. It is repelled from O by a force which is inversely proportional to x.

a) Express the situation as a second order differential equation connecting x, t and a constant k.

b) Modify your equation to include a resistive force proportional to the velocity of the particle.

5 A particle of mass 2 kg is at rest at a point O on a smooth horizontal plane. The particle is attached to O by a light elastic string of natural length 3 m and modulus of elasticity 20 N. The particle is given an impulse so that it moves away from O with initial speed 10 m s^{-1}.

M2

a) Write the equation of motion of the particle, after the string becomes taut, as a differential equation connecting its displacement x m from O and the time t s since the string became taut.

b) Modify your equation for the situation where the plane is rough with coefficient of friction 0.1.

6 A particle of mass m lies on a smooth plane inclined at 30° to the horizontal. It is attached to a point O at the top of the plane by means of an elastic string of natural length l and modulus of elasticity λ. The particle is held at rest at a point A on the plane, where A is directly down the plane from O and OA = l. The particle is released from rest. Write its equation of motion as a differential equation in terms of x, its displacement down the plane from A.

7.3 Solving first-order equations

The simplest differential equations to solve are of the form you met in Chapter 1, namely $\frac{dv}{dt} = f(t)$, $\frac{dx}{dt} = f(t)$ or $\frac{d^2x}{dt^2} = f(t)$. Such equations can be solved directly by integration (twice in the case of the second-order equation). The resulting relationship will contain an arbitrary constant of integration (two in the case of the second-order equation), and is called the **general solution** of the equation. It defines a whole family of solutions which differ only in the value of the constant(s).

Example 3

Find the general solution of $\frac{d^2x}{dt^2} = 4 - 12t$.

Integrating once, you obtain

$\frac{dx}{dt} = \int(4 - 12t)\,dt = 4t - 6t^2 + a$, where a is an arbitrary constant.

Integrating again, you obtain

$x = 2t^2 - 2t^3 + at + b$, where b is an arbitrary constant.

This is the general solution of the equation.

Separation of variables

For first-order differential equation which cannot be solved directly by integration as above, it may be possible to **separate the variables**. That is, if the equation is in terms of x and t, it can be rewritten in the form

$$g(x)\frac{dx}{dt} = f(t)$$

You then integrate both sides with respect to t, which gives

$$\int g(x)\frac{dx}{dt}\,dt = \int f(t)\,dt$$

The left-hand side of this can be simplified to give

$$\int g(x)dx = \int f(t)\,dt$$

and in practice it is usual to move directly to this without the intervening statement.

Once the equation is written in this form, you can solve the differential equation provided that you are able to integrate the functions f and g. The resulting relationship between x and t contains an arbitrary constant of integration, and is the general solution of the equation.

The equations you will meet in the context of this syllabus will all be soluble by separation of variables.

Example 4

Find the general solution of the equation $\frac{dx}{dt} = xt$.

First, you separate the variables to give

$$\frac{1}{x}\frac{dx}{dt} = t$$

Next you integrate, which gives

$$\int \frac{1}{x}\,dx = \int t\,dt$$

and hence $\ln x = \frac{1}{2}t^2 + C$

You only need one constant of integration. Had you introduced one for each integral, you could have subtracted one of them from both sides and simplified to give the result shown.

M2

This is the general solution of the equation. However, it is usual to rewrite it to give x in terms of t where this can easily be done.

You could write

$$x = e^{\frac{1}{2}t^2}e^C$$

giving the general solution as

$$x = Ae^{\frac{1}{2}t^2}$$

where $A = e^C$.

Of course, you may have to call on your knowledge of various integration techniques when solving problems.

Example 5

Find the general solution of the equation $\dfrac{dx}{dt} = (x+1)(x-2)$.

M2

First, you separate the variables to give

$$\frac{1}{(x+1)(x-2)}\frac{dx}{dt} = 1$$

You then integrate, giving

$$\int \frac{1}{(x+1)(x-2)}\,dx = \int 1\,dt$$

To integrate the left-hand side, it is necessary to express it as partial fractions.

$$\tfrac{1}{3}\int \frac{1}{(x-2)} - \frac{1}{(x+1)}\,dx = \int 1\,dt$$

and hence $\tfrac{1}{3}[\ln(x-2) - \ln(x+1)] = t + C$

This can be written as

$$\ln\left(\frac{x-2}{x+1}\right) = 3t + K \quad \text{where } K = 3C$$

If necessary, you could make x the subject of this, as follows.

$$\frac{x-2}{x+1} = e^{3t+K} = Ae^{3t} \quad \text{where } A = e^K$$

and hence $x = \dfrac{2 + Ae^{3t}}{1 - Ae^{3t}}$

The particular solution

The general solution of a first-order differential equation contains an arbitrary constant. If you have further information about the problem, you can find the value of that constant. You then have the **particular solution** to the problem.

You find the value of the constant which is consistent with a known pair of values of the variables. Often the information is given in the form 'the value of x when $t = 0$ is...'.

These are usually called **initial conditions** if the stated value of the independent variable is zero, and more generally are called **boundary conditions**.

Example 6

Find the general solution of the equation $\frac{dx}{dt} = \frac{\cos t}{x}$.

Hence find the particular solution if the boundary conditions are $x = 2$ when $t = \frac{1}{6}\pi$.

First, you separate the variables.

$$x\frac{dx}{dt} = \cos t$$

Then you integrate, giving

$$\int x\,dx = \int \cos t\,dt$$

and hence $\frac{1}{2}x^2 = \sin t + C$

This is the general solution, which you could rearrange to give

$$x = \sqrt{2\sin t + K}$$

where $K = 2C$.

For the particular solution, you substitute $x = 2$ and $t = \frac{1}{6}\pi$, which gives

$$2 = \sqrt{1 + K}$$

and hence $K = 3$

The particular solution is, therefore, $x = \sqrt{2\sin t + 3}$.

M2

Exercise 7B

1 Find the general solution of the following differential equations.

a) $\frac{dv}{dt} = 6\sin 2t$ b) $\frac{d^2x}{dt^2} = \frac{3}{\sqrt{t}}$

2 Find the particular solution of the equation in question 1 a), given that $v = 5$ when $t = 0$.

3 Find the particular solution of the equation in question 1 b), given that $x = 4$ and $\frac{dx}{dt} = 6$ when $t = 0$.

4 By separating the variables, find the general solutions of the following differential equations.

a) $\frac{dx}{dt} = \frac{t}{x}$ b) $\frac{dx}{dt} = \frac{t}{x^2 - 2x}$ c) $\frac{dx}{dt} = (x-3)(t-2)$

d) $\frac{dv}{dt} = t(v^2 - 4)$ e) $t\frac{dv}{dt} + v^2 = 1$ f) $\frac{dv}{dt} - 2v = vt$

g) $t\tan x\frac{dx}{dt} = 1$ h) $(1 - \cos t)\frac{dx}{dt} = \sin t\cot x$

5 Find the particular solution of the equation $\frac{dx}{dt} = \frac{\sin t}{\cos x}$, given that $x = 0$ when $t = 0$.

6 Given that $\frac{dx}{dt} = e^{(x+t)}$ and that $x = 0$ when $t = 0$, show that $x = \ln\left(\frac{1}{2 - e^t}\right)$.

7 Given that $x^2\frac{dx}{dt} = 2e^{t-x}$, and that $x = 0$ when $t = 0$, show that $t = x + \ln(\frac{1}{2}x^2 - x + 1)$.

8 Use the substitution $y = x - t$ to rewrite $\frac{dx}{dt} = (x - t)^2$ as a differential equation in y and t. Hence

a) find the general solution of the original equation in the form $x = f(t)$

b) find the particular solution, given that $x = 0$ when $t = 0$

c) show that, in this case, $x = \frac{2}{1 + e^2}$ when $t = 1$.

M2

9 In Exercise 7A, question 1, you found a differential equation for the motion of an object falling from rest with resistance proportional to velocity and a terminal velocity of 60 m s^{-1}. By solving this equation, show that $v = 60(1 - e^{-\frac{gt}{60}})$, and hence find the time taken by the object to reach half its terminal velocity.

10 An object of mass 20 kg is moving at 30 m s^{-1} when it encounters a resistance of magnitude $5v$ N, where v is its speed.

a) Write the equation of motion of this object after it meets the resistance as a differential equation connecting v with t, the time which has elapsed since it encountered the resistance.

b) Find the particular solution of your equation.

c) Find the velocity of the object when $t = 2$.

d) Find the time which elapses before the speed of the object reduces to 5 m s^{-1}.

11 The combined mass of a cyclist and her machine is 100 kg. She starts from rest on a horizontal road against a resistance of magnitude kv, where v is her speed. She exerts a power of 1000 W and her maximum speed is 10 m s^{-1}.

a) Find the value of k.

b) Write her equation of motion as a differential equation in v and t.

c) Find the particular solution to your equation.

d) Hence find the time she takes to reach half her maximum speed.

Summary

You should know how to ...	Check out
1 Use $a = \frac{dv}{dt}$ to form differential equations.	**1** A particle is falling vertically through water. Its acceleration is given by $a = g - kv$, where k is a constant. Express this as a differential equation.
2 Solve differential equations to obtain relationships between velocity and time.	**2** Find the general solution for v for the differential equation found in question 1.
3 Use $v = \frac{dx}{dt}$ to form differential equations.	**3** A body moves so that its velocity is given by $v = 3\sin 2t + 4\cos 3t - e^t$. Express this as a differential equation.
4 Solve differential equations to obtain relationships between displacement and time.	**4** Find the general solution for x for the differential equation found in question 3.

Revision exercise 7

1 A possible model for the acceleration, a m s^{-2}, of a particle at time t seconds is

$$a = 8 - ht$$

where h is a positive constant.

a) The acceleration is zero when $t = 4$.
   i) Find h.
   ii) Write down an expression for a in terms of t.

b) The velocity of the particle is 2 m s^{-1} when $t = 4$. Find the velocity of the particle at time t. (*AQA, 2004*)

2 A racing car accelerates in a straight line from rest to a speed of 25 m s^{-1} in 2 seconds. The acceleration of the car, t seconds after the start, is a m s^{-2} and is modelled by the equation

$a = kt^2 + 10t$, where k is a constant and $0 \leqslant t \leqslant 2$.

The velocity of the car at time t seconds is v m s^{-1}.

a) i) Find v in terms of k and t.
   ii) Hence show that $k = \frac{15}{8}$.

b) Find the distance travelled by the car in the first 2 seconds of its motion. (*AQA, 2002*)

3 A parachutist, of mass 80 kg, is falling vertically. When his speed is 30 m s^{-1}, his parachute opens. He then experiences an air resistance force of magnitude $196v$ N, where v m s^{-1} is his speed.

a) Show that at time t seconds after the parachute is opened, the speed of the parachutist is given by

$$v = 4 + 26e^{-2.45t}$$

b) Sketch a graph to show how the parachutist's speed varies with time. (*AQA, 2001*)

4 A small sphere of mass m is fired into a tube of heavy oil, so that it initially moves vertically downwards at 10 m s^{-1}. The oil exerts a resistive force of magnitude $3mv$, where v is the speed of the sphere at time t.

a) Show that

$$\frac{dv}{dt} = 10 - 3v.$$

b) Show that $v = \dfrac{10 - Ae^{-3t}}{3}$ where A is a constant.
Find A and the terminal speed of the sphere.

c) Find the distance travelled by the sphere in the first 5 seconds. *(AQA, 1997)*

5 A car accelerates from rest along a straight horizontal road. It experiences a constant horizontal forward force of magnitude 2000 newtons and a resistance force. The resistance force has magnitude $40v$ newtons, when the speed of the car is $v\text{ m s}^{-1}$. The mass of the car is 1000 kg.

M2

a) Show that

$$\frac{dv}{dt} = \frac{50 - v}{25}$$

b) Find the velocity of the car at time t. *(AQA, 2003)*

6 A particle moves on a straight line. At time t seconds its acceleration, $a\text{ m s}^{-2}$, is given by

$$a = 20 \sin 4t$$

a) Initially the particle is at rest. Find an expression for the velocity of the particle at time t.

b) Initially the displacement of the particle from the origin is 0.8 m. Find an expression for the displacement of the particle at time t. *(AQA, 2004)*

7 A particle of mass m is moving along a straight horizontal line. At time t the particle has speed v. Initially the particle is at the origin and has speed U. As it moves the particle is subject to a resistance force of magnitude mkv^3.

a) Show that

$$v^2 = \frac{U^2}{2kU^2t + 1}.$$

b) What happens to v as t increases? *(AQA, 2003)*

8 A particle moves along a straight line. At time t seconds the acceleration, a m s^{-2}, of the particle is given by

$$a = 2 - 2e^{-t}$$

At time $t = 0$ the particle is at the origin moving with a velocity of 4 m s^{-1}.

a) Show that the velocity, v m s^{-1}, at time t seconds is given by

$$v = 2t + 2e^{-t} + 2$$

b) Find an expression for the distance of the particle from the origin at time t. *(AQA, 2003)*

9 A particle, of mass m, moves in a straight line on a smooth horizontal surface. As it moves it experiences a resistance force of magnitude kv^2, where k is a constant and v is the speed of the particle at time t. The particle moves with speed U at time $t = 0$.

Show that $v = \dfrac{mU}{Ukt + m}$. *(AQA, 2002)*

M2 Practice Paper (Option A)

75 minutes *60 marks* *You may use a graphics calculator.*

Answer ***all*** *questions.*

1 The orbit of the Moon around the Earth may be modelled as a circular. The time taken for the Moon to make one complete orbit is approximately 27.32 days.

a) Show that the angular speed of the Moon is approximately 2.66×10^{-6} radians per second. *(3 marks)*

b) Assuming that the radius of the circular orbit is approximately 3.844×10^{8} metres, find the speed of the Moon, relative to the Earth, in metres per second. *(2 marks)*

2 The diagram shows a uniform lamina which consists of two rectangular parts, ABCD and PQRS.

AB = CD = 2 cm. BC = AD = SR = PQ = 20 cm and RQ = SP = 10 cm.

The mid-point, O, of CB coincides with the mid-point of SP.

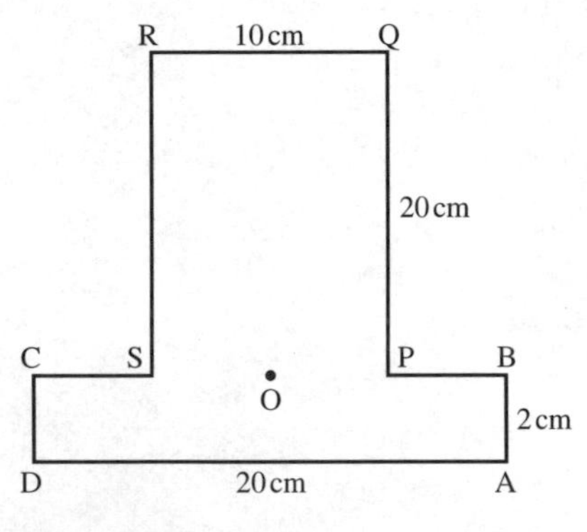

a) Find the distance of the centre of mass of the lamina from RQ. *(4 marks)*

b) The lamina is suspended from R. Find the angle that RQ makes with the vertical through R. *(3 marks)*

3 An elastic string has modulus of elasticity 20 N and natural length 0.5 metres. A particle of mass 0.4 kg is attached to one end of the string. The other end of the string is attached to a fixed point P. The particle is pulled down until it is 2 metres below P.

a) Calculate the elastic potential energy of the string when the particle is 2 metres below P. *(2 marks)*

b) The particle is released.

i) Show that the kinetic energy of the particle is 39.12 J when the string becomes slack. *(2 marks)*

ii) Find the kinetic energy of the particle when it is 0.5 metres below P. *(2 marks)*

iii) Find the maximum height of the particle above P. *(6 marks)*

4 A lorry of mass 6000 kg moves in a straight line along a horizontal road. During this motion the power of the lorry's engine is 30 kilowatts.

When the lorry is travelling at a speed of $10\ \text{m s}^{-1}$, the resistance to the motion of the lorry is of magnitude 1800 N. At this speed find

a) the tractive force produced by the engine *(2 marks)*

b) the acceleration of the lorry. *(2 marks)*

5 A particle has mass 500 kg. A single force, $\mathbf{F} = 2000t\mathbf{i} - 4000\mathbf{j}$ newtons, acts on the particle, at time t seconds. No other forces act on the particle.

a) Find the acceleration of the particle. *(2 marks)*

b) At time $t = 0$, the velocity of the particle is $8\mathbf{i}\ \text{m s}^{-1}$. Show that the velocity, $\mathbf{v}\ \text{m s}^{-1}$, of the particle at time t is

$$\mathbf{v} = (8 + 2t^2)\mathbf{i} - 8t\mathbf{j}$$ *(4 marks)*

c) Initially the particle is at the origin. Find the position vector, $\mathbf{r}$ metres, of the particle at time t. *(4 marks)*

6 A stuntman S stands initially on a platform P. He holds a rope which is attached to a fixed support O. The points O and P are at the same horizontal level. The man then steps off the platform with the rope taut and horizontal. He swings through a circular arc, as shown in the diagram.

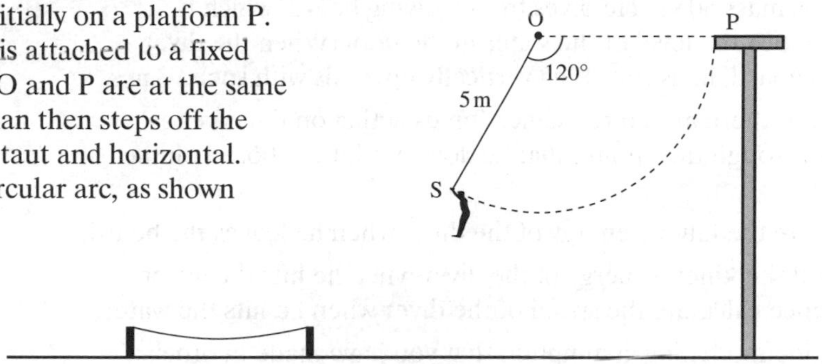

O P 120° 5 m S

The mass of the man is 70 kg and the length of the rope is 5 m. The man may be modelled as a particle throughout the motion.

a) The man releases his grip on the rope when it has swung through an angle of 120°.

i) Show that the speed of the man when he is about to let go of the rope is approximately $9.21\ \text{m s}^{-1}$. *(4 marks)*

ii) Find the tension in the rope when the man is about to let go of the rope. *(5 marks)*

b) After letting go of the rope, the stuntman moves freely under gravity.

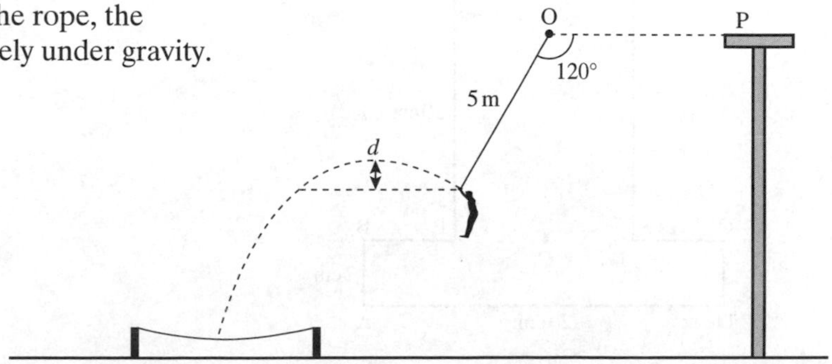

O P 120° 5 m d

He rises a distance d before beginning to fall into the safety net. Show that d is approximately 1 metre. *(4 marks)*

c) Comment on the assumption that the man can be modelled as a particle. *(1 mark)*

7 A stone, of mass m, falls vertically under gravity through still water.

As the stone falls, it experiences a resistance force of magnitude λmv N.

a) If the speed of the stone at time t is v, show that

$$\frac{dv}{dt} = g - \lambda v$$ *(2 marks)*

b) If the initial speed of the stone is u, show that

$$v = \frac{g}{\lambda} - \frac{(g - \lambda u)e - \lambda t}{\lambda}$$ *(6 marks)*

M2 Practice Paper (Option B)

90 minutes *75 marks* *You may use a graphics calculator.*

Answer ***all*** *questions.*

1 A diver has mass 50 kg. He dives from a diving board, which is 4 metres above the level of the water in the pool. When the diver leaves the board, he is travelling vertically upwards with speed 3 m s^{-1}.

Assume that there are no resistance forces acting on the diver as he moves through the air and that he does not hit the board on the way down.

a) Calculate the kinetic energy of the diver when he leaves the board. *(2 marks)*

b) i) Find the kinetic energy of the diver when he hits the water. *(3 marks)*

ii) Hence calculate the speed of the diver when he hits the water. *(2 marks)*

c) State one modelling assumption that you have made in order to answer the question. *(1 mark)*

2 The diagram shows a uniform lamina which consists of two rectangular parts, ABCD and PQRS.

AB = CD = 2 cm. BC = AD = SR = PQ = 20 cm and RQ = SP = 10 cm.

The mid-point, O, of CB coincides with the mid-point of SP.

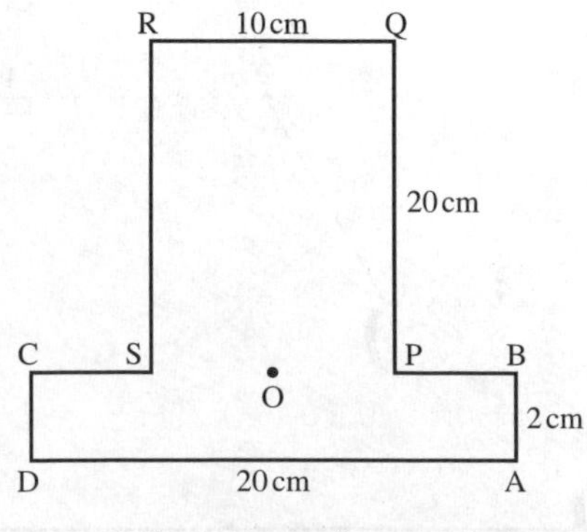

a) Find the distance of the centre of mass of the lamina from RQ. *(4 marks)*

b) The lamina is suspended from R. Find the angle that RQ makes with the vertical through R. *(3 marks)*

3 A particle moves along a straight line. At time t the displacement of the particle from its initial position is x where $x = 4t^2 + 4e^{-2t} + 7$

a) Find, at time t:
   i) the velocity of the particle *(2 marks)*
   ii) the acceleration of the particle *(2 marks)*

b) Describe what happens to the acceleration of the particle as t increases. *(2 marks)*

4 An elastic string has modulus of elasticity 20 N and natural length 0.5 metres. A particle of mass 0.4 kg is attached to one end of the string. The other end of the string is attached to a fixed point P. The particle is pulled down until it is 2 metres below P.

a) Calculate the elastic potential energy of the string when the particle is 2 metres below P. *(2 marks)*

b) The particle is released.
   i) Show that the kinetic energy of the particle is 39.12 J when the string becomes slack. *(2 marks)*
   ii) Find the kinetic energy of the particle when it is 0.5 metres below P. *(2 marks)*
   iii) Find the maximum height of the particle above P. *(6 marks)*

5 A lorry of mass 6000 kg moves in a straight line along a horizontal road. During this motion the power of the lorry's engine is 30 kilowatts. When the lorry is travelling at a speed of 10 m s^{-1}, the resistance to the motion of the lorry is of magnitude 1800 N. At this speed find

a) the tractive force produced by the engine *(2 marks)*

b) the acceleration of the lorry. *(2 marks)*

6 A particle has mass 500 kg. A single force, $\mathbf{F} = 2000t\mathbf{i} - 4000\mathbf{j}$ newtons, acts on the particle, at time t seconds. No other forces act on the particle.

a) Find the acceleration of the particle. *(2 marks)*

b) At time $t = 0$, the velocity of the particle is $8\mathbf{i}$ m s^{-1}. Show that the velocity, $\mathbf{v}$ m s^{-1}, of the particle at time t is
   $$\mathbf{v} = (8 + 2t^2)\mathbf{i} - 8t\mathbf{j}$$ *(4 marks)*

c) Initially the particle is at the origin. Find the position vector, $\mathbf{r}$ metres, of the particle at time t. *(4 marks)*

7 A ladder, of mass 15 kg, leans against a smooth wall at an angle of 70° to the horizontal. The base of the ladder is on rough horizontal ground. The coefficient of friction between the ground and the ladder is μ.

a) Find the force exerted by the wall on the ladder. *(2 marks)*

b) Find an inequality that μ must satisfy. *(4 marks)*

8 A stuntman S stands initially on a platform P. He holds a rope which is attached to a fixed support O. The points O and P are at the same horizontal level. The man then steps off the platform with the rope taut and horizontal. He swings through a circular arc, as shown in the diagram.

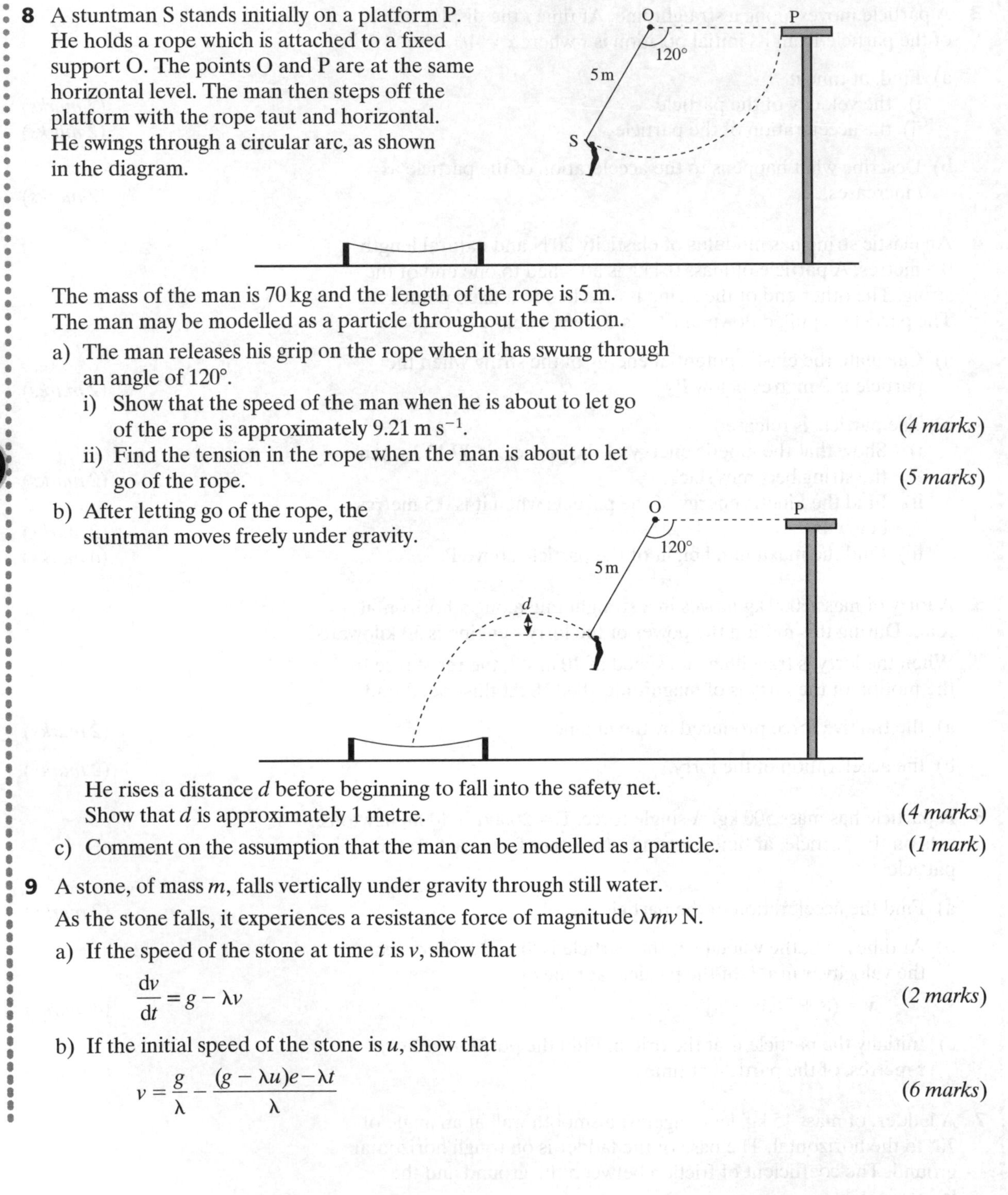

The mass of the man is 70 kg and the length of the rope is 5 m. The man may be modelled as a particle throughout the motion.

a) The man releases his grip on the rope when it has swung through an angle of 120°.

i) Show that the speed of the man when he is about to let go of the rope is approximately 9.21 m s^{-1}. *(4 marks)*

ii) Find the tension in the rope when the man is about to let go of the rope. *(5 marks)*

b) After letting go of the rope, the stuntman moves freely under gravity.

He rises a distance d before beginning to fall into the safety net. Show that d is approximately 1 metre. *(4 marks)*

c) Comment on the assumption that the man can be modelled as a particle. *(1 mark)*

9 A stone, of mass m, falls vertically under gravity through still water. As the stone falls, it experiences a resistance force of magnitude λmv N.

a) If the speed of the stone at time t is v, show that

$$\frac{\mathrm{d}v}{\mathrm{d}t} = g - \lambda v$$

(2 marks)

b) If the initial speed of the stone is u, show that

$$v = \frac{g}{\lambda} - \frac{(g - \lambda u)e^{-\lambda t}}{\lambda}$$

(6 marks)

M2

8 Coursework guidance

This chapter is for students taking the MM2A unit.
The MM2B unit does not contain coursework.
If you are unsure which unit you are taking, you should ask your teacher.

In this chapter you will find:

- A clear definition of how to tackle your Mechanics coursework
- A strand by strand breakdown of the marking grid specifically geared to the A2 tasks
- Useful tips and hints from experienced moderators
- Answers to some frequently asked questions
- A final checklist.

M2

8.1 Introduction

This will probably be the second piece of Mechanics coursework that you attempt in your mathematics course. It should build on the experiences and skills that you developed when tackling the AS level tasks in the MM1A unit. If you have not yet attempted any mechanics coursework do not worry, as the guidance in this chapter will still help you to successfully complete your task.

The emphasis of the coursework will be to mathematically model a given 'real-life' situation. You will need to make and justify assumptions for your model, define suitable variables and constants, perform relevant calculations, draw graphs, interpret your results and attempt to validate them by comparing them to real-life examples.

The task will be assessed at A2 standard so there will be increased expectations of the work that you produce. For the highest marks there will need to be a maturity of expression, clarity in your use of mechanical principles and methods, as well as clear and appropriate discussion of your results in the context of the task. Use the skills you have developed in the unit to produce an interesting and relevant write-up which clearly displays your understanding of the mechanics.

This chapter will help you with the coursework process right from the starting point to handing in your completed piece of work. There are useful hints and tips from experienced moderators who work for the Examination Board, as well as a clear and full description of the marking grid which will be used to assess your piece of work.

For your MM2A coursework you will need to submit one task. This is worth 25% of the marks available for the unit ($4\frac{1}{6}$% of the total A-level award).

8.2 Choosing a task

The Examination Board for this specification will provide a list of tasks which are appropriate for your MM2A coursework. Your teacher may decide to offer one task or a number of tasks for you to choose from. It is important to choose a task that you feel comfortable with and one which gives you the scope to use your mechanics skills fully.

Listen carefully to the advice of your teacher. Do not start a task that has not been approved by your teacher.

Just as in your MM1A task your teacher may provide some time in class to discuss various ideas and approaches that you might take when tackling your task. Write down what you might do, discuss these in small groups, adapt and modify your ideas and reject as necessary. Think carefully about how the criteria in the marking grid will apply to your task and what you will need to do to address these criteria.

You could write your ideas as a series of bullet points.

This discussion process is particularly important in Mechanics, as the modelling process needs clear explanation to ensure that the correct and logical approach is taken.

M2

Most importantly learn from any mistakes you made in your approaches to the MM1A task. Your teacher will be able to help you with this.

The main focus of the coursework tasks for MM2A is modelling circular motion or modelling elasticity.

8.3 The modelling process

The modelling process can be broken down into a series of stages, illustrated by the flowchart:

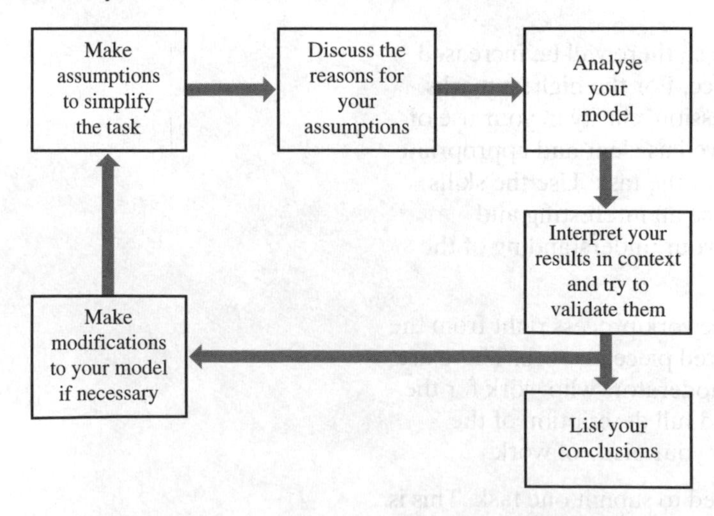

You should try to follow this process when conducting your coursework task.

8.4 Assessment criteria

Mechanics coursework is marked under four strands:

- ✦ Formulating the model
- ✦ Analysing the model
- ✦ Interpreting and validating the model
- ✦ Communication

There are 80 marks in total. This marking grid shows how the marks are allocated.

Strand	0–8 marks	9–15 marks	16–20 marks
1. Formulating the model	Problem defined and understood. Some simplifying assumptions made.	Assumptions stated and discussed and linked to a simple model. Appropriate choice of variables.	Comprehensive model set up. All assumptions clearly stated and discussed, where appropriate.
2. Analysing the model	Data collected and/or organised. Some accurate calculations and analysis of the problem.	Analyse the model using appropriate numerical and graphical or numerical and algebraic methods.	Analyse the comprehensive model. Clear and logical use of appropriate graphical or algebraic methods.
3. Interpreting and validating the model	Outcomes stated in everyday language. Some attempt to comment on the appropriateness of the numerical results.	Attempt to interpret and validate/or justify the solutions. Some limitations identified. Some consideration of refinements.	A reasoned attempt to interpret/validate/justify the solutions. Discussion of both limitations and refinements.
4. Communication	Some help given with the task if needed. Work clearly presented and organised.	Problem tackled with persistence and some initiative. Use of appropriate mathematical language or diagrams.	Coherent and logical approach to the task. Clear explanation of the findings/implications for further work.
			Total Mark Maximum 80

M2

It is important to remember that although the marking criteria are identical for all mechanics tasks, the criteria need to be interpreted in the context of the MM2A task you choose.

The following section discusses in detail the four strands that you will be assessed on. Reference is made to the marking grid to help you understand exactly what is expected of you.

Strand 1: Formulating the model

You need to create a mathematical model which you can analyse. This will require you to make assumptions, discuss the reasons for them fully and then define variables and constants you will use in your analysis.

Candidates describe what they are going to investigate and how they hope to do it. They identify those problems to be modelled mathematically. (4 marks)

Once you have chosen your task you need to decide how you are going to tackle it.

Question: What particular aspect of the task are you going to develop?

Do not state in your aims that you are going to do something and then not bother to do it!

A short introduction discussing what you are going to do and how you are going to do it is an excellent start. You can mention briefly which techniques you will be using and what extensions you will consider.

M2

Any background information you can use to help with the introduction adds to the readability of the piece.

If you were modelling a bungee jump you might look at the history of the activity as well as explaining to the reader what is involved in a bungee jump. To get this information you might use quoted sites from the Internet or reference books.

The piece of work will make all the necessary assumptions and discuss the reasons for them. (8 marks)

When you are looking to analyse a real-life situation, there are many different and complex factors that could potentially be included in your model. However, there is a danger that if you try to consider and incorporate everything that is relevant, you will end up with an over-complicated model which would be far too difficult to analyse.

If you were modelling a car travelling along a road, you could consider many factors such as:

- The mass of the car
- The size of the car
- Air resistance
- The contact between the wheels and the road
- Surface material of the road
- ✦ Weather conditions – appropriate value for the coefficient of friction μ.
- Is the road straight, on a bend or banked?
- The effect of gravity
- ✦ The engine

You need to decide which factors are important. You can then discuss what you are going to assume about these factors and what their role will be in your model. Some factors can be assumed to have no effect for **your** model.

Other students may include something you don't and vice versa. There is no 'perfect' correct answer.

In many situations, it is appropriate to treat relatively large moving bodies as point masses. This simplifies the problem considerably. It may be that you decide to ignore air resistance in your model. Try to explain why you are doing so in your own words.

Referring to the car example, you might decide that its mass is important. You could then assume that the car is a particle, meaning that it will have mass but no size.

You need to consider all aspects that may affect your model and make appropriate assumptions accordingly. It is vital that you do not just make a list of assumptions with no discussion of them. You **must** explain in your own words why you have made these assumptions.

As a guide, the list of assumptions and their explanation are worth 4 marks each.

All assumptions should be sensible with either practical collection of data or evidence of research for them, where appropriate. (6 marks)

The assumptions you have made should be sensible and appropriate.

If you are modelling the loop of a roller coaster, it would certainly not be sensible to assume that the loop is of height zero. Also you might assume that the roller coaster starts from rest, but is this sensible?

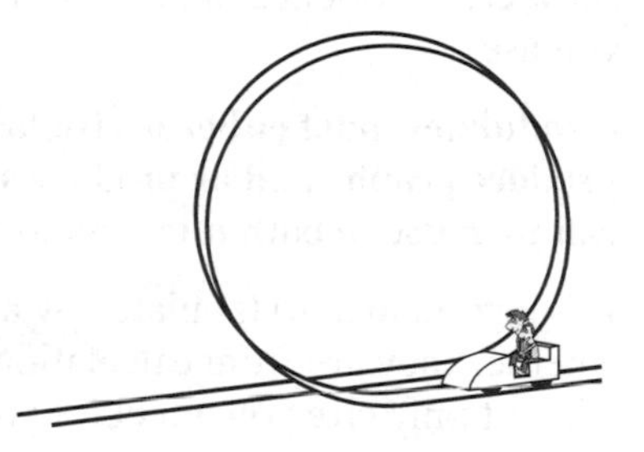

There should be clear evidence of any research you have used to make assumptions. There is a wide range of websites on the Internet which give specific details and values you could use in your coursework.

If you are modelling a bungee jump, you need values to use for the length of the rope as well as the stiffness of the rope to calculate the extension during the jump. You will probably need to obtain these from appropriate websites (whose addresses should be stated) as it would be difficult to carry out a sensible experiment to obtain these values for yourself.

If an experiment is used, the details should be given and the results clearly quoted. If a constant is needed, try to research its value rather than just guessing a value for it which is not appropriate.

Remember that the experiment is to support your model, not to become the focus of the task. Your coursework is not a write-up of an experiment; it is a mathematical model of a real-life situation.

Suitable variables and constants should be defined fully. (2 marks)

Any variables or constants you will use in your analysis should be defined: for example, v = velocity in m s^{-1}.

Units must be quoted for the full two marks.

Strand 2: Analysing the model

The calculations should be correct and appropriate to the model and content of the unit. (10 marks)

You will need to analyse the model that you have set up.

For the tasks at A2 level you will need to set up a generalisation of your model at a fairly early stage. The algebraic formulae that you suggest can then be graphed (from the formula or a table of values generated from it) and interpreted. You may also generate spreadsheets to enable you to vary one of the components of your formulae to further aid your interpretation.

Ensure that you have a sufficient range of calculations.

If you are analysing a roller coaster, you need to consider the conditions required to 'loop the loop' and also consider the 'g' forces acting on the passengers at different points on the loop. You may also decide to consider other loop shapes such as clothoids.

The calculations need to be accurate and in your write-up you should show clear evidence of at least one of each type of calculation that you use.

Candidates must either derive tables of values and use them to produce graphs and/or produce suitable algebraic formulae that they can then use to both interpret and interpolate. (8 marks)

It is a good idea to tabulate any numerical results that you obtain, whether they are from calculations or derived from a formula. This will not only give you a feel for your results, but will also flag up any obvious errors in your calculations.

You are expected as a minimum to either:

- Obtain a table of values from calculations and then produce a graph from them, or
- Obtain some generalised results and then interpret them.

The nature of the tasks at A2 level lend themselves more to an algebraic approach from which you can develop your interpretation and modifications.

Appropriate degrees of accuracy should have been used in the answers given. (2 marks)

Always think carefully about the answers that you give to ensure that they are appropriate for the values assumed earlier in your work. If you are consistent, you will obtain 1 mark. Work that is both fully appropriate and consistent will gain 2 marks.

Strand 3: Interpreting and validating the model

To address this strand you will look at your solutions and attempt to validate them. There is a tendency to spend too little time on this section, but it is worth the same amount of marks as the other three sections.

Candidates must predict what will happen to values of variables not calculated directly. These may be taken from a graph or formula and should be validated if possible. (4 marks)

You should be able to look at the formulae and graphs that you have obtained and make predictions from them.

It is useful to ask yourself the question: 'What would happen if ...?'

You might be able to make a prediction by:

- Changing a particular value in your general result.
- Using a graph that you have drawn (using a value not used to draw the graph in the first place).

When you predict from a graph you should use a value that was not used as part of the initial data.

M2

If you are modelling the motion of a car you might state that: 'For values of μ less than 0.3 the car would....'

They must look at how realistic their final results are, and, if appropriate, give reasons why the answers are not sensible. (6 marks)

You will have obtained a number of results.

Question: Are your results sensible and realistic?

To address this question, you could do one of the following:

- Carry out a practical experiment to test the results and predictions of the model.
- Compare your results with real-life examples of what you are modelling.
- Look for data on the Internet to confirm or contradict your results.

If you are modelling roller coasters there are many real-life examples in theme parks all over the world. You could also build a model of a roller coaster to test your results.

If you are modelling a car going round a bend you could consider banked roads or possibly banks on a motor racing circuit such as the Indianapolis-type circuits in America.

If you are modelling a bungee jump you could consider real-life jumps in places such as Australia and New Zealand.

Do not worry if your results are different from real-life examples. However, you should try to explain **why** your results are different. Is there any particular aspect or assumption that you made which could contribute to or be the major cause of the difference?

They must look at the effects the assumptions have. This process should be done for all of the assumptions which directly affect the answer. They must look at what modifications could be made to the model in order to get a more realistic result (but they should not have to make them unless their model is so over-simplified that this becomes necessary). (10 marks)

An important aspect of the interpretation of your model is to look back at your original assumptions and see what effects they had on the eventual solution that you obtained. For **each** individual assumption, discuss any difference there would be to your answer if that assumption had been changed.

It may be that you introduce something which had been ignored.

In modelling the time taken for an object to fall 50 metres to the ground, you may have assumed that air resistance was negligible. You could state: 'If air resistance had been included, the total time taken would have been increased because …'

M2

By making assumptions you usually make the task simpler. However, if your model is over-simplified it may become unworkable in reality.

You also need to suggest modifications that would provide a more realistic model. This is the opportunity to consider other aspects of the task which could have been looked at originally.

In modelling a roller coaster, you could have considered the friction between the coaster and the track. You would not need to consider this modification in detail unless your original model was so simple that the analysis became trivial.

Strand 4: Communication

The final strand measures how well you have communicated your work in the final write-up. This is a section which credits the good approaches used in the first three strands.

Candidates must express themselves clearly and concisely using appropriate mathematical language and notation. Graphs and diagrams should be clearly and accurately labelled. (4 marks)

You will be assessed on how clearly you have expressed your ideas overall. It is important to be concise.

You do not need to include page after page of repetitive calculations, but you must include at least one worked example of all types of calculation that you have used in your coursework.

It is important to use the appropriate mathematical terms and notation. It is expected that in a piece of mechanics there should be a force diagram drawn (if appropriate) with all relevant forces shown.

Candidates should include other areas of work which could have been investigated further. (2 marks)

You are expected to suggest other work which could have followed from your coursework.

Question: Where could you take your research further?

If you were modelling a roller coaster you could suggest considering other sorts of fairground rides which have different types of motion (with examples).

The final conclusions should be set out logically. (2 marks)

Your final conclusions should be stated clearly at the end of your coursework.

Question: What exactly did your analysis indicate?

The overall piece of work should be of sufficient depth and difficulty. (6 marks)

This section gives credit for all of the work that has been done throughout your coursework. Ensure that your approaches have enabled you to generate enough mathematics for you to have shown the appropriate skills from the unit.

You will receive credit not only for the difficulty of your calculations but for the quality and depth of your interpretation and the overall approach.

It is not expected that you will be looking for ways of making the coursework artificially difficult.

The investigation should form a coherent whole and should be of a length consistent with a piece of work of 8–10 hours. (6 marks)

Your report should read as a logical piece of writing that should be easy to follow.

It is not expected that you will use the strands or strand breakdowns as specific titles in your coursework, but from your writing it should be clear to the reader which strand is being addressed.

If you struggle to follow the flow of your arguments when reading the work through, then so will the person marking and assessing it. It is expected that your piece of work will take 8–10 hours to complete. This is an approximate timing including data collection and write-up time.

It is easy to get carried away with some tasks so be careful that your piece of work does not become too long.

When it is complete, read through your work and check your calculations to ensure that you have not made any careless mistakes.

8.5 Frequently asked questions

Here are some questions that are often asked by students.

Should the coursework be hand-written or word-processed?

It can be either. If handwritten, try to ensure that the work is neat and clear to follow. If word-processed, take care when typing symbols.

How long should the piece of work be?

An appropriate piece of work could vary from 10 sides up to 20 sides including diagrams. (Word-processed pieces tend to be shorter).

Should I label the page numbers?

Yes. It is useful when the work is being moderated.

Can I use the Internet?

Yes. The Internet is appropriate to collect relevant information your task, and it will be useful to obtain constants you may need for your model. **Always** quote any websites used.

Any attempt to copy work from the Internet is against Examination Board rules and could lead to serious consequences.

Make sure that you are aware of the deadline set by your teacher and work to it.

8.6 Checklist

M2

Have you ....?

Strand 1

- Stated your aims clearly
- Listed your assumptions and discussed the reasons for them
- Ensured your assumptions are sensible and listed any sources of information such as websites
- Provided clear details of any experiments used to collect data practically
- Defined all variables and constants with appropriate units

Strand 2

- Checked the accuracy of your calculations
- Ensured that you have a full range of appropriate calculations
- Ensured that you have produced tables of values and graphs or algebraic formulae and graphs
- Used appropriate degrees of accuracy in your answers

Strand 3

- Made predictions for values of variables not calculated directly
- Looked at how realistic your final results are and if not have you tried to give reasons
- Looked at the effect of all of the assumptions made on your results and suggested modifications that could be made to your model

Strand 4

- ✦ Expressed yourself clearly using clear diagrams and appropriate mathematical language and notation
- ✦ Suggested other areas of work that could have been considered
- ✦ Set out your final conclusions clearly
- ✦ Made sure your work is of sufficient depth and difficulty (have you made it too simplistic)
- ✦ Made sure your work is easy to follow and of sufficient length

Appendix

Principle of moments

You can find the line of action of the resultant of a pair of like parallel forces, as follows.

Suppose like forces P N and Q N act at points A and B. The resultant of the forces has a magnitude of $(P + Q)$ N and acts at some point C, as shown.

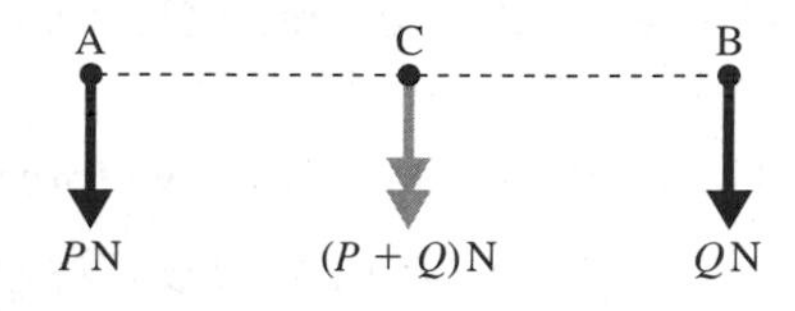

You can add a pair of equal and opposite forces to the system without changing the overall situation. Suppose you add forces of PQ N at A and B, as shown below.

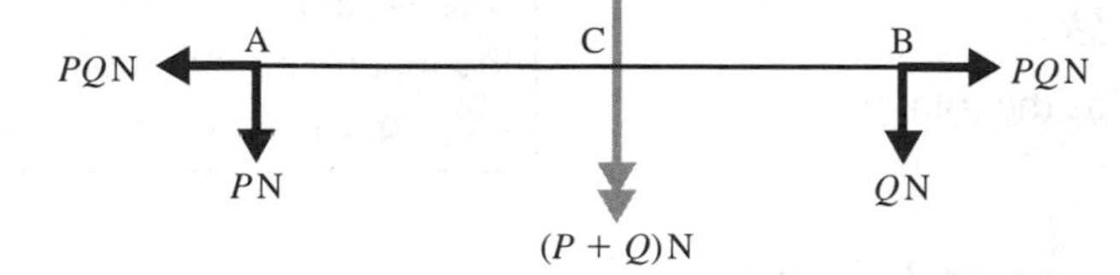

The forces which now act at A have a resultant of magnitude

$$\sqrt{P^2Q^2 + P^2} = P\sqrt{(Q^2 + 1)}\text{ N}$$

acting at an angle θ, where

$$\tan\theta = \frac{PQ}{P} = Q \text{ as shown.}$$

Similarly, the forces which now act at B have a resultant of magnitude

$$Q\sqrt{(P^2 + 1)}\text{ N}$$

acting at an angle ϕ, where $\tan\phi = P$, as shown.

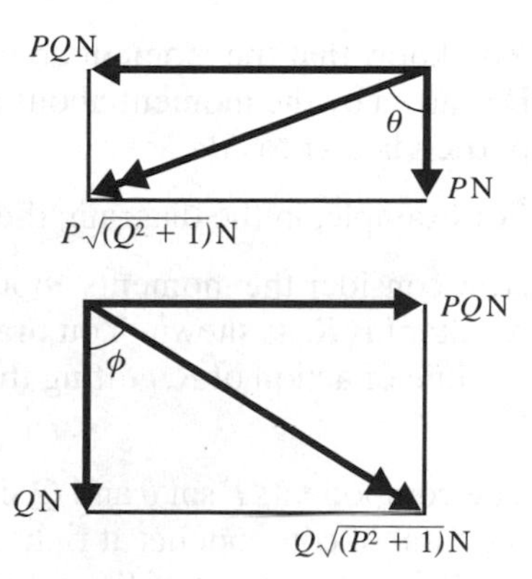

The pairs of forces at A and B can be replaced by these resultant forces. The lines of action of these forces intersect at D, as shown.

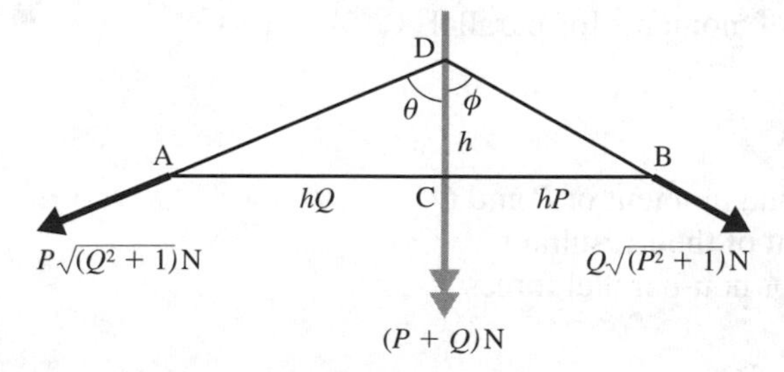

The resultant of the whole system is still $(P + Q)$ N and you can now see that its line of action must pass through D.

If the length CD = h, then you have

$$\tan\theta = Q \quad\Rightarrow\quad AC = hQ$$
$$\tan\phi = P \quad\Rightarrow\quad BC = hP$$

So, the line of action of the resultant divides the line AB internally in the ratio $Q : P$.

For **unlike** forces of magnitude P N and Q N, where $P > Q$, you can follow a similar procedure (you might like to try this for yourself).

The resultant is now a force of magnitude $(P - Q)$ N. Its line of action is through the point C, where AC : BC = $Q : P$ as before, but C now divides AB **externally** in the ratio $Q : P$.

The principle of moments for parallel forces

Suppose you have like forces of magnitude P N and Q N and any point O. The line through O at right angles to the direction of the forces cuts the lines of action of P and Q at A and B respectively, and the line of action of their resultant at C, as shown.

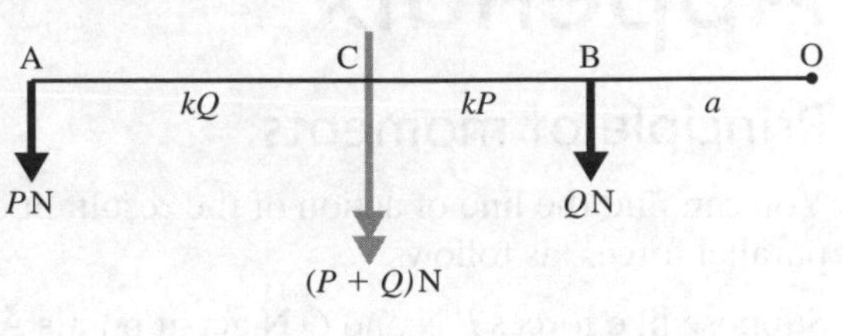

C divides AB in the ratio $Q : P$, so you can put AC = kQ and BC = kP. Let BO = a.

The total moment of the two forces about O is

$$P(kQ + kP + a) + Qa = kPQ + kP^2 + Pa + Qa$$

The moment of the resultant about O is

$$(P + Q)(kP + a) = kPQ + kP^2 + Pa + Qa$$

So, the moment of the resultant is the same as the total moment of the forces.

A similar argument will establish this for unlike parallel forces. Why not check this for yourself?

M2

Principle of moments for non-parallel forces

You know that the moment about point A of a force F acting at point B is given by the moment about A of the component of the force perpendicular to AB.

For example, in the diagram, the moment of F about A is $Fa \sin \theta$.

Now consider the moments about point O of forces P and Q whose resultant is R, as shown. You draw the perpendicular OA from O to the line of action of R, cutting the lines of action of P and Q at B and C.

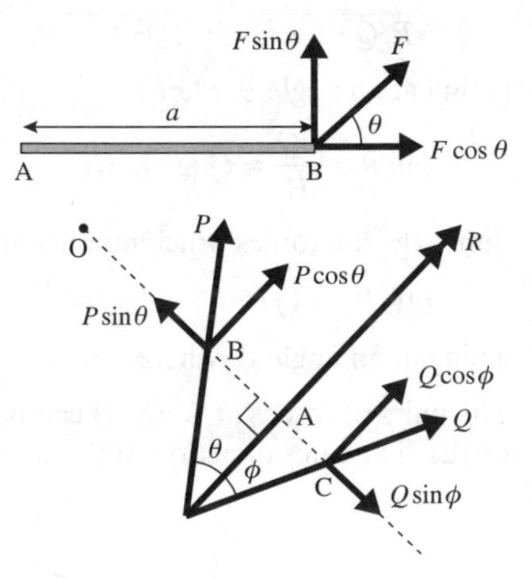

The components $P \sin \theta$ and $Q \sin \phi$ are equal and opposite (because the total of components at right angles to the resultant must be zero). So, R is the resultant of the components $P \cos \theta$ and $Q \cos \phi$. These are parallel forces, therefore, by the principle of moments for parallel forces,

$$P \cos \theta \times \text{OB} + Q \cos \phi \times \text{OC} = R \times \text{OA}$$

But the left-hand side of this equation is the total moment of P and Q about O, and the right-hand side is the moment of their resultant about O. So, the principle of moments holds for non-parallel forces.

Centre of mass v centre of gravity

For a very large object, the centre of gravity would not usually coincide with the centre of mass.

Consider, for example, a system consisting of two equal masses m placed at A and B as shown, where A is twice as far as B from E, the centre of the Earth.

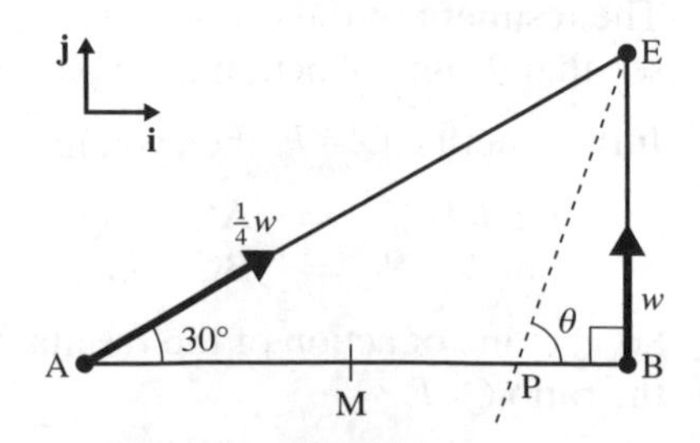

The centre of mass of the system is at M, the mid-point of AB.

However, by the inverse square law for gravity, the weight of particle A is a quarter of that of particle B and it acts in a different direction.

The resultant of the two weights is

$$\mathbf{R} = \tfrac{1}{4}w\cos 30°\mathbf{i} + (w + \tfrac{1}{4}w\sin 30°)\mathbf{j}$$
$$= \frac{w\sqrt{3}}{8}\mathbf{i} + \frac{9w}{8}\mathbf{j}$$

The line of action of $\mathbf{R}$ is PE, where $\tan\theta = \dfrac{9}{\sqrt{3}}$.

This means that $BP = \dfrac{BE\sqrt{3}}{9}$. But $AB = BE\sqrt{3}$ and so the resultant weight does not act through the centre of mass.

In fact, the centre of gravity of this system would be that point on PE at which a mass of $2m$ would have a weight of $\mathbf{R}$. This centre of gravity would move in relation to A and B if AB were moved in relation to E.

M2

Concurrence of medians

To prove that the medians of a triangle are concurrent, and that their common point divides each median in the ratio 2 : 1.

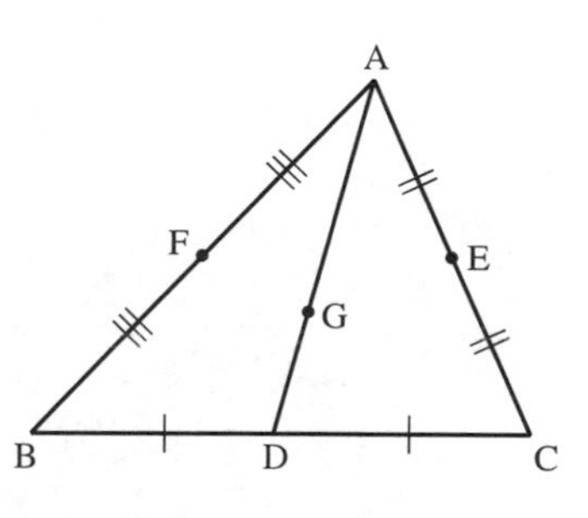

Let the position vectors of A, B and C relative to some origin be $\mathbf{a}$, $\mathbf{b}$ and $\mathbf{c}$.

The mid-point, D, of BC has position vector

$$\mathbf{d} = \mathbf{b} + \tfrac{1}{2}\overrightarrow{BC} = \mathbf{b} + \tfrac{1}{2}(\mathbf{c} - \mathbf{b}) = \tfrac{1}{2}(\mathbf{b} + \mathbf{c})$$

By a similar argument the mid-points E and F of AC and AB have position vectors

$$\mathbf{e} = \tfrac{1}{2}(\mathbf{a} + \mathbf{c}) \quad \text{and} \quad \mathbf{f} = \tfrac{1}{2}(\mathbf{a} + \mathbf{b})$$

Now consider G, the point on AD such that AG : GD = 2 : 1.
Its position vector is

$$\mathbf{g} = \mathbf{a} + \tfrac{2}{3}\overrightarrow{AD}$$
$$= \mathbf{a} + \tfrac{2}{3}(\mathbf{d} - \mathbf{a})$$
$$= \mathbf{a} + \tfrac{2}{3}(\tfrac{1}{2}\mathbf{b} + \tfrac{1}{2}\mathbf{c} - \mathbf{a})$$
$$= \tfrac{1}{3}(\mathbf{a} + \mathbf{b} + \mathbf{c})$$

But the point which divides BE in the ratio 2 : 1 has position vector

$$\mathbf{b} + \tfrac{2}{3}(\mathbf{e} - \mathbf{b}) \quad \text{which simplifies to} \quad \tfrac{1}{3}(\mathbf{a} + \mathbf{b} + \mathbf{c})$$

and the point which divides CF in the ratio 2 : 1 has position vector

$$\mathbf{c} + \tfrac{2}{3}(\mathbf{f} - \mathbf{c}) \quad \text{which simplifies to} \quad \tfrac{1}{3}(\mathbf{a} + \mathbf{b} + \mathbf{c})$$

Hence the point G lies on all three medians and divides each in the ratio 2 : 1.

Work done by a variable force

A small change, δF, in the force will accompany a small change, δx, in the displacement. During this change, the force lies between F and $(F + \delta F)$, and hence the work done, δW, lies between $F\delta x$ and $(F + \delta F)\delta x$. You have

$$F\delta x \leqslant \delta W \leqslant (F + \delta F)\delta x$$

and so $F \leqslant \dfrac{\delta W}{\delta x} \leqslant (F + \delta F)$

As $\delta x \to 0$, $\delta F \to 0$ and $\dfrac{\delta W}{\delta x} \to \dfrac{dW}{dx}$.

So, in the limit, you have $F \leqslant \dfrac{dW}{dx} \leqslant F$ and hence $\dfrac{dW}{dx} = F$

It follows that $W = \int F\,dx$, and as x changes from a to b, you have

M2

$$W = \int_a^b F\,dx$$

Answers

Chapter 1

Check in

1 a) $9x^2 + 2$ b) $\frac{5}{2}x^{\frac{3}{2}} - \frac{7}{2x^{\frac{3}{2}}} - \frac{6}{x^3}$ c) $\frac{2}{2x+1}$ d) $5e^{5x}$ e) $4\cos 4x$ f) $3x^2\cos 2x - 2x^3\sin 2x$

2 a) $\frac{1}{5}x^5 + \frac{1}{2}x^4 + c$ b) $\frac{4}{3}x^{\frac{3}{2}} + 10\sqrt{x} + c$ c) $\frac{1}{5}e^{5x} + c$ d) $-\frac{\cos 3x}{3} + c$ **3** a) $\frac{1}{5}(x+5)^5 - \frac{5}{4}(x+5)^4 + c$ b) $\frac{1}{3}e^{x^3} + c$ **4** $20\frac{2}{3}$

5 a) $\sqrt{160}$ or $4\sqrt{10}$ b) $27\mathbf{i} - \mathbf{j}$ **6** $\mathbf{v} = \mathbf{u} + \mathbf{a}t$, $r = \mathbf{u}t + \frac{1}{2}\mathbf{a}t^2$, $\mathbf{r} = \frac{1}{2}(\mathbf{u} + \mathbf{v})t$, $\mathbf{r} = \mathbf{v}t - \frac{1}{2}\mathbf{a}t^2$ **7** $(\mathbf{i} + 2\mathbf{j})\ \mathrm{m\,s^{-2}}$

Exercise 1A

1 a) $30t - 3t^2$, $15t^2 - t^3$ b) $75\ \mathrm{m\,s^{-1}}$, 250 m c) 500 m d) 15 s **2** a) −2 m b) 6 m

3 a) 0 s, 4 s, −4 s (that is, when timing starts and 4 s before and after this b) $\pm\frac{4}{\sqrt{3}}$ s (before and after timing starts)

c) $\pm\frac{128}{3\sqrt{3}}$ m d) $30\ \mathrm{m\,s^{-2}}$ **4** a) $2t^2 + 6t - 8$ b) −4, 1 c) 4 s before timing started

5 a) −1 s, 1 s, 2 s b) $6\ \mathrm{m\,s^{-1}}$, $-2\ \mathrm{m\,s^{-1}}$, $3\ \mathrm{m\,s^{-1}}$, $-10\ \mathrm{m\,s^{-2}}$, $2\ \mathrm{m\,s^{-2}}$, $8\ \mathrm{m\,s^{-2}}$ **6** a) $t^2 - 5t + 6$ b) 2 s and 3 s c) $7\frac{1}{3}$ m and $7\frac{1}{6}$ m

7 a) 30 s b) 225 m c) model requires acceleration and deceleration at start and end to be instantaneous

8 a) 2 s b) $v = 8 - 10t$ c) $-12\ \mathrm{m\,s^{-1}}$ d) 7.2 m e) Gives displacement (height above the ground) not distance travelled

9 a) 6, 8.2, 6.8, 6.3, 6.1, 6.0 b) −6, −2.2, −0.8, −0.3, −0.1, 0 c) 6, 2.2, 0.8, 0.3, 0.1, 0

d) Model implies that the body never reaches complete rest **10** a) $v = -60 + \frac{120}{t+1}$, $s = -60t + 120\ln(t+1)$

10 c) Approaches constant velocity $-60\ \mathrm{m\,s^{-1}}$ **11** $110\ \mathrm{m\,s^{-1}}$ **12** a) 6 s b) 72 m **13** $v = 2\ln(t+1)$, 19.1 s

Exercise 1B

1 a) $4\mathbf{i} + 4t\mathbf{j}$, $4\mathbf{j}$ b) $2(t-2)\mathbf{i} + (3t^2 - 4t)\mathbf{j}$, $2i + (6t-4)\mathbf{j}$

c) $-2\sin t\mathbf{i} + 2\cos t\mathbf{j} + \frac{1}{2\sqrt{t}}\mathbf{k}$, $-2\cos t\mathbf{i} - 2\sin t\mathbf{j} - \frac{1}{4t^{\frac{3}{2}}}\mathbf{k}$ d) $e^t\mathbf{i} + \frac{1}{t+1}\mathbf{j}$, $e^t\mathbf{i} - \frac{1}{(t+1)^2}\mathbf{j}$

2 a) $\mathbf{r} = \frac{1}{4}t^4\mathbf{i} + t^3\mathbf{j} + \mathbf{c}$, $\mathbf{a} = 3t^2\mathbf{i} + 6t\mathbf{j}$ b) $\mathbf{r} = 15t\mathbf{i} + (20t - 5t^2)\mathbf{j} + \mathbf{c}$, $\mathbf{a} = -10\mathbf{j}$ c) $\mathbf{r} = \left(\frac{1}{2}t^2 - \frac{1}{3}t^3\right)\mathbf{i} + \left(\frac{3}{2}t^2 - 5t\right)\mathbf{j} + \mathbf{c}$, $\mathbf{a} - (1-2t)\mathbf{i} + 3\mathbf{j}$

d) $\mathbf{r} = (t^2 - 2t^3)\mathbf{i} + (t^3 - t^4)\mathbf{j} + \mathbf{c}$, $\mathbf{a} = (2 - 12t)\mathbf{i} + (6t - 12t^2)\mathbf{j}$ e) $\mathbf{r} = 2t^2\mathbf{i} + 4\sin t\mathbf{j} - 2\cos t\mathbf{k} + \mathbf{c}$, $\mathbf{a} = 4\mathbf{i} - 4\sin t\mathbf{j} + 2\cos t\mathbf{k}$

f) $\mathbf{r} = 8\cos t\mathbf{i} + 8\sin t\mathbf{j} + 2e^{2t}\mathbf{k} + \mathbf{c}$, $a = -8\cos t\mathbf{i} - 8\sin t\mathbf{j} + 8e^{2t}\mathbf{k}$ **3** a) $\mathbf{v} = 3t^2\mathbf{i} + (2t - 2t^2)\mathbf{j} + \mathbf{c}$, $\mathbf{r} = t^3\mathbf{i} + (t^2 - \frac{2}{3}t^3)\mathbf{j} + \mathbf{c}t + \mathbf{d}$

3 b) $\mathbf{v} = -10t\mathbf{j} + \mathbf{c}$, $\mathbf{r} = -5t^2\mathbf{j} + \mathbf{c}t + \mathbf{d}$ c) $\mathbf{v} = 2t\mathbf{i} + 2t^2\mathbf{j} + \mathbf{c}$, $\mathbf{r} = t^2\mathbf{i} + \frac{2}{3}t^3\mathbf{j} + \mathbf{c}t + \mathbf{d}$

d) $\mathbf{v} = -5\sin t\mathbf{i} + 5\cos t\mathbf{j} + 3t^2\mathbf{k} + \mathbf{c}$, $\mathbf{r} = 5\cos t\mathbf{i} + 5\sin t\mathbf{j} + t^3\mathbf{k} + \mathbf{c}t + \mathbf{d}$ **4** a) $2(t^2+1)\mathbf{i} + (2t^4 - 1)\mathbf{j}$

4 b) $(7 - 4\cos t)\mathbf{i} + 2(t+1)\mathbf{j}$ c) $5(\cos t - 1)\mathbf{i} + (-1 - 5\sin t)\mathbf{j} + \frac{1}{2}(3\sin 2t - 10)\mathbf{k}$

5 a) $(3t - t^2 + 3)\mathbf{i} + \frac{1}{2}t^2(2 - 3t^2)\mathbf{j}$, $\frac{1}{6}(9t^2 - 2t^3 + 18t + 6)\mathbf{i} + \frac{1}{30}(10t^3 - 9t^5 - 60)\mathbf{j}$,

b) $2(1 + \sin 2t)\mathbf{i} + (5 - 4\cos 2t)\mathbf{j}$, $(2t - \cos 2t - 1)\mathbf{i} + (5t - 2\sin 2t + 4)\mathbf{j}$,

c) $(2t^2 - 3t + 2)\mathbf{i} - 3\mathbf{j} + t(3t - 2)\mathbf{k}$, $\frac{1}{6}(4t^3 - 9t^2 + 12t + 6)\mathbf{i} + (1 - 3t)\mathbf{j} + (t^3 - t^2 + 2)\mathbf{k}$ **6** a) $3\mathbf{i} + 4(1-t)\mathbf{j}$, $-4\mathbf{j}$ b) $5\ \mathrm{m\,s^{-1}}$

6 c) $t = 1$ s d) No. Always an x component of velocity. **7** $7\ \mathrm{m\,s^{-1}}$ **8** a) $2(t^2 - t + 1)\mathbf{i} + (t^3 + 3)\mathbf{j}$ b) $4\mathbf{i} + 9\mathbf{j}$

9 a) $2\mathbf{i} + 2(2-t)\mathbf{j}$, $2(t+1)\mathbf{i} + (4t - t^2 - 3)\mathbf{j}$ b) 2 s c) 1 s or 3 s **10** a) $480(\mathbf{i} + \sqrt{3}\mathbf{j})\ \mathrm{km\,h^{-1}}$ b) $(480t\mathbf{i} + 480\sqrt{3}t\mathbf{j} + 0.8\mathbf{k})$ km

11 $\mathbf{v} = (2t^2 - 2)\mathbf{i} + (12 - 3t)\mathbf{j}$, $\mathbf{r} = (\frac{2}{3}t^3 - 2t + 2)\mathbf{i} + (3 + 12t - \frac{3}{2}t^2)\mathbf{j}$. When $t = 6$, $\mathbf{v} = 70\mathbf{i} - 6\mathbf{j}$, $\mathbf{r} = 134\mathbf{i} + 21\mathbf{j}$

12 Speed $6\ \mathrm{m\,s^{-1}}$, $\mathbf{r} = 9\mathbf{i} - 4.5\mathbf{j} - 9\mathbf{k}$ **13** a) $(3 + 16t)\mathbf{i} + (1 + 24t)\mathbf{j}$ b) $16\mathbf{i} + 24\mathbf{j}$ c) 2 s or $\frac{2}{7}$ s

13 d) $40\mathbf{i} + 56\mathbf{j}$ away, $1.80\mathbf{i} + 6.69\mathbf{j}$ towards

Check out

1 $v = t^3 + \frac{1}{2}e^{2t} - \frac{1}{2}$, $s = \frac{1}{4}t^4 + \frac{1}{4}e^{2t} - \frac{1}{2}t$ **2** $\mathbf{v} = 6t\mathbf{i} + 3e^{3t}\mathbf{j} - 4\cos 4t\mathbf{k}$, $\mathbf{a} = 6t\mathbf{i} + 9e^{3t}\mathbf{j} + 16\sin 4t\mathbf{k}$ **3** $a = (t^2 + 8\sin 2t)\ \mathrm{m\,s^{-2}}$

Revision exercise 1

1 a) $-4\sin t\mathbf{i} + (3.5 - 3\cos t)\mathbf{j} + \frac{1}{2}t\mathbf{k}$ b) $4\cos t\mathbf{i} + (3.5t - 3\sin t)\mathbf{j} + \frac{1}{4}t^2\mathbf{k}$ c) 2.57 m **2** a) $t^2\mathbf{i} + (4t + 2)\mathbf{j}$ b) $12\mathbf{i} + 8\mathbf{j}$
3 a) $4\cos t\mathbf{i} - 4\sin t\mathbf{j} + 6\mathbf{k}$ b) $-4\sin t\mathbf{i} - 4\cos t\mathbf{j}$ d) $v^2 = 52$ **4** a) $4 - 2e^{-t}$
4 b) $2e^{-t}$ c) Acceleration decreases approaching a value of zero. **5** a) $(4t^3 - 4t)\mathbf{i} + (12t^2 - 4t^3)\mathbf{j}$ b) $(t^3 - t)\mathbf{i} + (3t^2 - t^3)\mathbf{j}$
5 c) $(3t^2 - 1)\mathbf{i} + (6t - 3t^2)\mathbf{j}$ d) $\frac{1}{\sqrt{3}}$ **6** a) $(3t^2 - 3)\mathbf{i} + (12t - 12)\mathbf{j}$ b) $(t^2 - 1)\mathbf{i} + (4t - 4)\mathbf{j}$ d) 5 **7** a) $\frac{1}{2}t\mathbf{i} - \frac{5}{2}\mathbf{j}$
7 c) $\frac{1}{12}t^3\mathbf{i} + (6t - \frac{5}{4}t^2)\mathbf{j}$ **8** a) i) 75 ii) 10 iii) 0 b) $h = 10, k = 25$ **9** a) $6\mathbf{i} + 2t\mathbf{j}$ b) $\sqrt{36 + 4t^2}$ c) 3
10 a) $(2t - 6)\mathbf{i} + t^2\mathbf{j}$ b) 3 c) $2\mathbf{i} + 2t\mathbf{j}$ **a** is a function of t and thus not constant. **11** a) 2 b) i) $4\mathbf{i} + 12\mathbf{j}$ ii) $-2\mathbf{i} - 4\mathbf{j}$
11 c) 16.5 **12** a) $\begin{pmatrix} 4t \\ 5 \end{pmatrix}$ b) 0.5 c) i) $\begin{pmatrix} 4 \\ 2t \end{pmatrix}$ ii) $\begin{pmatrix} 40 \\ 80 \end{pmatrix}$ **13** a) $\begin{pmatrix} t^2 - 4t \\ 3t + 2 \end{pmatrix}$ b) $\begin{pmatrix} 8 \\ 1 \end{pmatrix}$

Chapter 2

Check in

1 $(6.55\mathbf{i} + 4.59\mathbf{j})$ N **2** $2.5g$

Exercise 2A

1 a) +56 N m b) −87.5 N m c) −100.8 N m d) −31.5 N m e) +451.2 m f) +720 N m
2 a) i) +19 N m ii) −35.6 N m b) i) −18 N m ii) −24.4 N m c) i) −9.35 N m ii) −12.81 N m
3 a) +14.4 N m b) −9.6 N m c) +2.4 N m d) −4.8 N m **4** a) 1.1 m b) 0.6 m c) 0.8 m

Exercise 2B

1 a) +23.8 N m b) −26.1 N m c) −81.4 N m d) −152 N m e) +169 N m f) 324 N m **2** a) i) +7.89 N m ii) −60.4 Nm
2 b) i) −10.2 N m ii) −17.0 N m c) i) +9.43 N m ii) −9.99 N m **3** 24.0 N m **4** −12 N m **5** −17 N m **6** −8 N m

Exercise 2C

1 a) Resultant 8 N, 1.5 m from A, 2.5 m from B b) Couple c) Resultant 3 N, $1\frac{1}{3}$ m from A, $2\frac{2}{3}$ m from B
d) Resultant 1 N at mid-point of AB e) Equilibrium f) Couple **2** a) 8 N, 1 m b) 6 N, 0.5 m **3** a) 50 N, 150 N
3 b) 0.2 m **4** $46g$ N, $64g$ N **5** $a < x < 10a$ **6** $\frac{W(a + c - 2b)}{2c}, \frac{W(2b + c - a)}{2c}$ **7** a) $2\frac{2}{3}$ m b) 12.5 kg
9 26.6°. Swapping $3W$ and $2W$ gives equilibrium with AC at 33.7°

Exercise 2D

1 a) 63.7 N b) 73.5 N c) 127 N **2** a) Taking moments about B gives $T = 2W$, but resolving up needs $T = 1.5W$
b) $P = \frac{2W\sqrt{3}}{3}, AD = \frac{a}{3}$ **3** $0.258a$ **4** $P = 21.2$ N, 61.2 N at A, 42.4 N at B **5** a) 34.6 N b) 2.33 m **6** 63.4°
7 61.9° **8** $P = 29.7$ N, $\mu = 0.152$ **10** Horizontal $0.2W$, vertical $0.6W$

Check out

1 30.6 Nm **2** No resultant force, no resultant moment. **3** 17.3 N

Revision exercise 2

1 490 N, 196 N **2** $18g$ N, $22g$ N **3** a) $41\frac{2}{3}g$ N, $208\frac{1}{3}g$ N b) $\frac{25}{16}$ m **4** c) 374 N **5** b) ii) 3.71 **6** b) 67.3°
7 b) $\frac{5}{18}W$

Chapter 3

Check in

1 28 N m, 7 N

Exercise 3A

1 a) $(3\frac{4}{15}, 5\frac{8}{15})$ b) (1.3, 3.15) c) $(\frac{4}{11}, -6\frac{7}{11})$ d) $(2\frac{7}{11}, 0, -1\frac{8}{11})$ **2** $4\frac{7}{16}\mathbf{j} - 2\frac{3}{8}\mathbf{j}$ **3** $-\frac{5}{18}\mathbf{i} - 3\frac{8}{9}\mathbf{j} + \frac{2}{9}\mathbf{k}$ **4** (−13.6, −8.4)
5 $m = 4$ kg, CE = 1 m **6** $(2\frac{18}{19}, 2\frac{1}{19}, \frac{18}{19})$ **7** 48° **8** 46.4° **9** 2.79 kg

Exercise 3B

1 a) $(2\frac{5}{16}, 1\frac{11}{16})$ m b) (1.54, 0.641) m c) (0.923, 0.8) m d) (−0.049, 0) m e) (3.20, 1.81) m f) (59.7, 67.8) m
2 (0.578, 0.356) m **3** (41.1, 17.8) cm **4** (0.629, 0.793, 0.214) m **5** a) $(\frac{2}{5}, \frac{1}{3})$ m b) 31.0° **6** 21.8°
7 a) 86.9 cm b) 4.72 N down on handle, 11.1 N up on shaft **8** a) (29.0, 25, 60.8) cm b) 10.5°

Check out

1 10 cm **2** $6\frac{1}{9}$ cm, $3\frac{1}{3}$ cm **3** 8.31 cm **4** 25.7°

Revision exercise 3

1 b) 4.4 c) 63.4° **2** a) i) 2.6 ii) 28.9° b) 3 **3** a) 0.72 m b) symmetry in line through centre parallel to AD
3 c) 50.2° **4** a) Symmetrical about AP b) 10.96 cm c) 32° **5** b) 19.7° **6** a) i) 28.4 ii) 41.3 b) 28.4 c) 24.8
7 a) 5 cm c) 1

Chapter 4

Check in

1 $2\sqrt{gh}$ **2** a) $\sqrt{189}$ b) $\sqrt{2034}$ **3** 1.20, −9.20
4 $v = u + at, s = \frac{1}{2}(u + v)t, s = ut + \frac{1}{2}at^2, s = vt - \frac{1}{2}at^2, v^2 = u^2 + 2as$

M2

Exercise 4A

1 13 720 J **2** a) 1760 J b) 6760 J **3** a) 221 J b) $0.25gn(n-1)$ J **4** 1060 J **5** 33 N, 528 J **6** 2880 J
7 a) 2250 J b) 3720 J **8** 2780 J **9** a) 1020 m, 102 kJ b) 1250 m, 108 kJ c) 557 m, 52.4 kJ **10** 36 J **11** 23 J
12 34 J

Exercise 4B

1 15.3 m s^{-1}, mass not needed **2** a) 8.85 m s^{-1} b) 530 J, $88\frac{1}{3}$ N **3** a) 6.32 m s^{-1} b) 5.95 m c) 7.75 m s^{-1}
4 4.70 m s^{-1} **5** a) 3.43 m s^{-1} b) jerk as string went taut **6** a) $\sqrt{2ga\cos\theta}$ b) $\sqrt{2ga}$ **7** $\sqrt{\frac{2ga}{3}}$ **8** $\sqrt{2}:1$

Exercise 4C

1 a) 39.2 W b) $6\frac{8}{15}$ s **2** 2250 W **3** 6.28 W **4** 25.0 W **5** a) $1998a$ J b) 0.25 m s^{-1} **6** $5\frac{1}{6}$ W **7** 3.83 m
8 500 N

Exercise 4D

1 2 MW **2** 14 500 W **3** a) 10 m s^{-1} b) 0.625 m s^{-2} **4** a) 200 N b) 8.45 m s^{-1} or 30.4 km h^{-1} **5** 20.6 m s^{-1}
6 30 m s^{-1} **7** −0.030 m s^{-2}, 9.05 m s^{-1}, 0.255 m s^{-2} **8** $216\frac{2}{3}$ kW, $\frac{6500}{3(40 + 5n + 5g + ng)}$ m s^{-1} **9** a) 11 200 W
9 b) 0.773 m s^{-2}, 284 N **10** 0.221 m s^{-2} **11** a) 2520 N b) 22.3 m s^{-1} **13** 837 m **14** 5.2 m s^{-1} **15** $8P$ W

Check out

1 8000 J, 7250 J **2** $80g$ or 784 J decrease **3** 37.5 J **4** 9.35 m s^{-1} **5** 8.06 m s^{-1} **6** 2700 W

Revision exercise 4

1 a) i) 130 J ii) 0.204 m b) i) 3950 J ii) 11.0 m s^{-1} **2** a) 48.6 J b) 16.5 m c) 42.7 J, 16.9 ms^{-1}
3 a) i) 14.7 J ii) 10.8 ms^{-1} b) 4.73 N **4** b) 29.6 m s^{-2} **5** b) 7.53° **6** b) ii) 44 775 W c) 20.1 m s^{-1}
7 a) i) 1102.5 J ii) 7 m s^{-1} b) 8.93 m, 6.41 m s^{-1} c) 4.26 m **8** b) 21 c) 39.6 m s^{-1}

Chapter 5

Check in

1 No resultant force, no resultant moment.
2 The total mechanical energy of a system is constant unless external forces act or there are sudden changes (impacts etc).
3 The work done on a system by external forces equals the change in its energy. **4** $\frac{\lambda x^2}{2l}$ **5** 32.8 N

Exercise 5A

1 2.87 m **2** 50 N, 0.25 m **3** a) 0.27 m b) 44.1 N **4** a) 1.73 m b) 36.6 N, 12.4 N **5** 1.66 m, 0.745 m
6 a) 10.2 kg b) 13.2 kg **7** a) 200 N m^{-1} b) 1.1 m **8** 0.84 m, 100 N m^{-1} **9** 2.63 ms^{-2} **10** 30.4 cm
11 0.170 m **12** a) 2.31 m b) i) 1.70 m ii) 10.6 N, 9.04 N **13** 2.23 m **14** AM 2.99 m, BM 1.01 m **15** 3.46 m
16 a) 2.89 m, 2.11 m b) 3.78 m **17** $d = \frac{mgl}{2\lambda}$ **18** a) $\frac{mg\mu}{2l}$ b) At C (unrealistic for real springs)

Exercise 5B

1 a) 7.67 m s^{-1} b) $4\frac{1}{6}$ m **2** AB = 6.46 m, AC = 4.54 m **3** 3.26 m s^{-1} **4** 28.5 m s^{-1} **5** 24 J **6** a) 1.1 m
6 b) 0.6 m c) 0.2 m above A **7** a) $2\frac{8}{11}$ m b) Both 10.9 N c) 5.45 J d) AB 18.4 N tension, BC 9.09 N compression
7 e) 12.3 J f) 1.66 m s^{-1} g) 0.5 m **8** $2\sqrt{\frac{2ag}{3}}$ and $\sqrt{\frac{2ag}{3}}$. Collide $2a$ from start point of lighter particle **9** a) 0.4 m
9 b) 3.46 m s^{-1} **10** a) 3.97 m b) 8.82 m s^{-1}

Check out

1 $\frac{10}{3}mg$ N **2** $\frac{10}{3}mga$ J

Revision exercise 5

1 a) 8 J c) ii) Block stops 32.0 cm from the wall **2** a) i) 78.4 J ii) 8.85 m s^{-1} b) ii) 7.95 m
2 c) No air resistance, rope is light **3** a) 20 b) 47.2 c) i) 21.7 m s^{-1} **4** b) ii) 4.51 m
5 a) 16.6 m s^{-1} b) ii) 14 iii) −29.4 m s^{-2}

M2

Chapter 6

Check in

1 a) $3mg$ N **2** 10.7 ms^{-1} **3** a) $\frac{\pi}{4}$ b) 30° **4** a) $-12\sin 4t\,\mathbf{i} + 12\cos 4t\,\mathbf{j} - 6\mathbf{k}$ b) $-48\cos 4t\,\mathbf{i} - 48\sin 4t\,\mathbf{j}$
5 $(4\mathbf{i} + 4\mathbf{j} - 3\mathbf{k})$ m s^{-2} **6** $(5.44\mathbf{i} + 2.54\mathbf{j})$ N

Exercise 6A

1 a) $\frac{\pi}{2}$ rad s^{-1} b) $\frac{750}{\pi}$ rev min^{-1} **2** a) 9 m s^{-1} b) $\frac{5\pi^2}{12}$ m s^{-1} **3** a) 2.69 rad s^{-1} b) 4.85 m s^{-1} **4** 14.3 rad s^{-1}
5 1.02×10^{-4} = m s^{-1}, 1.92×10^{-3} m s^{-1}

Exercise 6B

1 25.6 m s^{-2} b) $66\frac{2}{3}$ m s^{-2} c) 0.5 m s^{-2} d) 37.8 m s^{-2} e) 2.5 m s^{-2} f) 2.05 m $^{-2}$
2 a) 7.27×10^{-5} rad s^{-1} b) 0.034 m s^{-2} c) 53.8° d) Perpendicular to Earth's axis of rotation
3 5.91×10^{-5} m s^{-2} **4** a) 0.6 rad s^{-1} b) $\mathbf{r} = (10\cos 0.6t\mathbf{i} + 10\sin 0.6t\mathbf{j})$ m
4 c) $\mathbf{v} = (-6\sin 0.6t\mathbf{i} + 6\cos 0.6t\mathbf{j})$ m s^{-1}, $\mathbf{a} = (-3.6\cos 0.6t\mathbf{i} - 3.6\sin 0.6t\mathbf{j})$ m s^{-2}
5 a) $\mathbf{a} = (-30\cos 10t\mathbf{i} - 30\sin 10t\mathbf{j})$ m s^{-2} b) $|\mathbf{a}| = 30$ m s^{-2} c) $(1.52\mathbf{i} + 2.59\mathbf{j})$ m s^{-1}

Exercise 6C

1 10.4 N **2** 213 N **3** 149 rev min^{-1} **4** a) 0.06 b) 0.081 c) Same in each case

5 a) 12 m s^{-1} b) 8.49 m s^{-1} **6** 0.48 N **7** $\dfrac{\mu g}{16\pi^2}$ **8** 5.07 rad s^{-1} **9** 3ω, $9ma\omega^2$

10 $\sqrt{\dfrac{g(1-\mu)}{a}} \leqslant \omega \leqslant \sqrt{\dfrac{g(1+\mu)}{a}}$ **11** a) 7.27×10^{-5} rad s^{-1} b) 42 200 km c) 1 h 49 min **12** $\sqrt{g}$ rad s^{-1}

Exercise 6D

1 0.392 m b) Ball is particle, string light and inextensible, no air resistance **2** 7 rad s^{-1} **3** 1.12 m, 7.40 N

4 b) AO > 0 for all finite values of ω. If string horizontal, no vertical component of tension to oppose particle's weight.

5 a) 2.01 N b) 1.68 m s^{-1} **6** a) 8.20 m s^{-1}, $6\frac{11}{18}m$ N b) 18.8 m s^{-1}, $16\frac{1}{3}m$ N **7** $3\frac{1}{3}$ m **8** 3.89 rads^{-1}

9 0.055 N at 27° to vertical **10** a) $\frac{3}{4}\omega^2 - g\sqrt{3}$ N b) 8.24 rads^{-1}

Exercise 6E

1 a) i) 6.29 m s^{-1} ii) 4.76 m s^{-1} iii) 2.38 m s^{-1} b) i) 96.2 N tension ii) 45.3 N tension iii) 5.62 N thrust

2 a) 8.57 m s^{-1} b) i) 4.41 N ii) 8.49 m s^{-1} **3** 48.2° **4** 70.5° **6** $\sqrt{8ag}$ **7** a) $U \geqslant 2\sqrt{ag}$ b) $U \geqslant 5\sqrt{ag}$

8 $\sqrt{\frac{1}{2}ag(3\sqrt{3}-4)}$ **9** 0.6 m **10** $\frac{1}{4}g\sqrt{51}$ N at 42.7° to vertical **11** 4.5 m **12** $\frac{1}{2}r$

13 183.75 N. Assumes mass is a particle, string light and inextensible, no air resistance.

14 $R_A = \frac{1}{3}mg(7 - 3\sqrt{3})$, $R_B = \frac{1}{3}mg(11 - 6\sqrt{3})$ **15** 10.1 m

M2

Check out

1 $\dfrac{150}{\pi}$ rev min^{-1} **2** a) 5 m b) 180 m s^{-2} towards the centre of the circle.

3 a) i) $-15 \sin 3t\mathbf{i} + 15 \cos 3t\mathbf{j}$ ii) $-45 \cos 3t\mathbf{i} - 45 \sin 3t\mathbf{j}$ b) (0, 0, −2) **4** a) $6g$ b) 3.13 rad s^{-1}

5 a) $\sqrt{\dfrac{4}{3}gr}$ b) $\dfrac{19mg}{3}$

Revision exercise 6

1 b) 90 cm c) $\dfrac{\pi}{5}$ m s^{-2} towards O **2** c) 4.9 rad s^{-1} **3** a) 160 m s^{-2} b) 22.6 N c) 309 N **4** c) $0.1k$ e) 0.003 J

5 a) i) 33.9 N b) 77° **6** b) i) $\frac{11}{15}mg$ ii) $\frac{23}{30}mg$ **7** b) $\sqrt{ag\sqrt{3}}$ **8** b) $\dfrac{u^2 + 2gl}{3gl}$

10 a) $\sqrt{10ag}$ b) ii) $\dfrac{10mga}{d} - 5mg$ c) Ball is assumed to be a particle, or no air resistance or no jolt at P.

11 b) ii) 34° iii) 530 N iv) inelastic string; no air resistance; modelled girl as a particle **12** a) $ml\omega^2$ c) ii) 86°

13 a) $2gr$ b) iii) $mg(2 + 3\sin\theta)$ d) $-\frac{1}{2}g$

Chapter 7

Check in

1 $p = \dfrac{mg}{m-1}$ **2** a) $\frac{1}{5}t^5 + \frac{1}{2}t^4 + c$ b) $\frac{2}{3}t^{\frac{3}{2}} - \frac{3}{t} + c$ c) $\frac{1}{7}e^{7t} + c$ d) $-\frac{1}{2}\cos 4t + c$

3 a) $\frac{1}{5}(t-2)^5 + \frac{1}{2}(t-2)^4 + c$ b) $3\ln(t-2) - \ln(t+4) + c$ c) $-\frac{1}{3}te^{-3t} + \frac{1}{9}e^{-3t} + c$ d) $t\ln t - t + c$

Exercise 7A

1 $5\dfrac{dv}{dt} = 5g - kv$, $k = \dfrac{49}{60} = \dfrac{g}{12}$ **2** $\dfrac{dv}{dt} = k\sin\pi t$, $k = 4$, $t = 0.5$ **3** $500\dfrac{dv}{dt} = \dfrac{5000}{v} - kv^2$

4 a) $m\dfrac{d^2x}{dt^2} = \dfrac{k}{x}$ b) $m\dfrac{d^2x}{dt^2} = \dfrac{k}{x} - K\dfrac{dx}{dt}$ **5** a) $\dfrac{d^2x}{dt^2} = -\dfrac{10x}{3}$

5 b) $\dfrac{d^2x}{dt^2} = -\dfrac{10x}{3} \pm 0.1g$ where the sign is + if $\dfrac{dx}{dt} < 0$ and − if $\dfrac{dx}{dt} > 0$ **6** $\frac{1}{2}mg - \dfrac{\lambda x}{l} = m\dfrac{d^2x}{dt^2}$

Exercise 7B

1 a) $v = c - 3\cos 2t$ b) $x = 4t^{\frac{3}{2}} + ct + d$ **2** $v = 8 - 3\cos 2t$ **3** $x = 4t^{\frac{3}{2}} + 6t + 4$

4 a) $x^2 = t^2 + k$ b) $2x^3 - 6x^2 = 3t^2 + k$ c) $x = Ae^{\frac{1}{2}(t-2)^3} + 3$ or $x = Ae^{\frac{1}{2}t^2 - 2t} + 3$ d) $v = \dfrac{2(1 + A^{2t^2})}{(1 - Ae^{2t^2})}$ e) $v = \dfrac{At^2 - 1}{At^2 + 1}$

f) $v = Ae^{\frac{1}{2}(t+2)^2}$ g) $\sec x = At$ h) $\sec x = A(1 - \cos t)$ **5** $\sin x = 1 - \cos t$ **8** a) $x = t + \dfrac{1 + Ae^{2t}}{1 - A^{2t}}$ b) $x = t + \dfrac{1 - e^{2t}}{1 + e^{2t}}$

9 4.24 s **10** a) $4\dfrac{dv}{dt} = -v$ b) $v = 30e^{-\frac{1}{4}t}$ c) $18.2\ \text{m s}^{-1}$ d) 7.17 s **11** a) 10 b) $\dfrac{100}{v} - v = 10\dfrac{dv}{dt}$

11 c) $t = 5\ln\left(\dfrac{100}{100 - v^2}\right)$ or $v = 10\sqrt{1 - e^{-\frac{1}{5}t}}$ d) 1.44 s

Check out

1 $\dfrac{dv}{dt} = g - kv$ **2** $v = \dfrac{1}{k}(g + ce^{-kt})$ **3** $\dfrac{dx}{dt} = 3\sin 2t + 4\cos 3t - e^t$ **4** $x = -\frac{3}{2}\cos 2t + \frac{4}{3}\sin 3t - e^t + c$

Revision exercise 7

1 a) i) 2 ii) $8 - 2t$ b) $8t - t^2 - 14$ **2** a) i) $v = \frac{k}{3}t^3 + 5t^2$ b) $15\frac{5}{6}$ m

4 b) $A = -20$ speed is $\frac{10}{3}\ \text{m s}^{-1}$ c) $\left(\frac{170}{9} - \frac{20}{9}e^{-15}\right)$ m **5** b) $50(1 - e^{-\frac{t}{25}})\ \text{m s}^{-1}$

6 a) $(5 - 5\cos 4t)$ b) $(5t - \frac{5}{4}\sin 4t + 0.8)$ **7** b) v tends to zero **8** $t^2 - 2e^{-t} + 2t + 2$

M2

M2 Practice Paper A (Option A)

1 b) $1020\ \text{m s}^{-1}$ **2** a) $\dfrac{71}{6}$ b) 67.1° **3** a) 45 J b) i) 39.12 J ii) 35.2 J iii) 1.23 m **4** a) 3000 N b) $0.2\ \text{m s}^{-2}$

5 a) $4t\mathbf{i} - 8\mathbf{j}$ c) $(8t + \frac{2}{3}t^3)\mathbf{i} - 4t^2\mathbf{j}$ **6** a) ii) 1780 N

6 c) Height of man is significant compared to length of the rope. Air resistance would reduce the speed.

M2 Practice Paper B (Option B)

1 a) 225 J b) i) 2185 J ii) $9.35\ \text{m s}^{-1}$ c) Diver is a particle. **2** a) $\dfrac{71}{6}$ b) 67.1°

3 a) $8t - 8e^{-2t}$ b) $8 + 16e^{-2t}$ c) reduces to the value of 8 **4** a) 45 J b) i) 39.12 J ii) 35.2 J iii) 1.23 m

5 a) 3000 N b) $0.2\ \text{m s}^{-2}$ **6** a) $4t\,\mathbf{i} - 8\,\mathbf{j}$ c) $(8t + \frac{2}{3}t^3)\,\mathbf{i} - 4t^2\,\mathbf{j}$ **7** 26.8 N b) $\mu \geq 0.182$

8 a) ii) 1780 N c) Height of man is significant compared to length of the rope. Air resistance would reduce the speed.

Formulae

You should learn the following formulae, which are **not** included in the AQA formulae booklet, but which may be required to answer questions.

Centre of mass	$\bar{X}\sum m_i = \sum m_i x_i$ and $\bar{Y}\sum m_i = \sum m_i y_i$
Circular motion	$v = r\omega,\ a = r\omega^2$ and $a = \dfrac{v^2}{r}$
Work and energy	Work done, constant force: Work $= Fd\cos\theta$
	Work done, variable force: Work $= \int F\,\mathrm{d}x$
	Gravitational Potential Energy $= mgh$
	Kinetic Energy $= \frac{1}{2}mv^2$
	Elastic Potential energy $= \dfrac{\lambda}{2l}e^2$
Hooke's law	$T = \dfrac{\lambda}{l}e$

The following formulae, introduced in M1, will also need to be remembered as they will be required to answer questions in M2.

Constant Acceleration Formulae	$s = ut + \frac{1}{2}at^2$	$\mathbf{s} = \mathbf{u}t + \frac{1}{2}\mathbf{a}t^2$
	$s = vt - \frac{1}{2}at^2$	$\mathbf{s} = \mathbf{v}t - \frac{1}{2}\mathbf{a}t^2$
	$v = ut + at$	$\mathbf{v} = \mathbf{u} + \mathbf{a}t$
	$s = \frac{1}{2}(u + v)t$	$\mathbf{s} = \frac{1}{2}(\mathbf{u} + \mathbf{v})t$
	$v^2 = u^2 + 2as$	
Weight	$W = mg$	
Momentum	Momentum $= mv$	
Newton's Second Law	$F = ma$ or Force = Rate of change of momentum	
Friction, dynamic	$F = \mu N$	
Friction, static	$F \leqslant \mu N$	

Index